Botulinum Toxin - Recent Topics and Applications

Edited by Suna Sabuncuoglu

Published in London, United Kingdom

Botulinum Toxin - Recent Topics and Applications
http://dx.doi.org/10.5772/intechopen.98126
Edited by Suna Sabuncuoglu

Contributors
Anna V. Reznik, Francisca del Rosario González Núñez, Francisco Núñez de Castro, Rosa Maria María González Núñez, Maria Angelo-Khattar, Mohamed H. Zahran, Ali Abdel Raheem, Ibrahim Alowidah, Diaa-Eldin Taha, Giovanni Palleschi, Antonio Cardi, Imad Katbeh, Mohammad Osama Makkeiah, Tamara Kosyreva, Lada Saneeva

First published in London, United Kingdom, 2022 by IntechOpen
IntechOpen is the global imprint of INTECHOPEN LIMITED, registered in England and Wales, registration number: 11086078, 5 Princes Gate Court, London, SW7 2QJ, United Kingdom

British Library Cataloguing-in-Publication Data
A catalogue record for this book is available from the British Library

Additional hard and PDF copies can be obtained from orders@intechopen.com

Botulinum Toxin - Recent Topics and Applications
Edited by Suna Sabuncuoglu
p. cm.
Print ISBN 978-1-80355-228-6
Online ISBN 978-1-80355-229-3
eBook (PDF) ISBN 978-1-80355-230-9

Meet the editor

Suna Sabuncuoglu graduated from the Faculty of Pharmacy, University of Hacettepe, Ankara, Turkey. She completed her Ph.D. studies in Pharmaceutical Toxicology and worked as a Ph.D. student at the International Agency for Research on Cancer, Molecular Carcinogenesis Laboratory. She also held a post-doc position at the Department of Chemotherapy and Virology, Rega Institute, The Catholic University of Leuven, Belgium. Dr. Sabuncuoğlu became a lecturer in 2013 and an associate professor in 2014. To date, she has served on many different boards, commissions and centers in or out of the university. Since 2018, she has held the title of European Registered Toxicologist (ERT). She continues to work as an associate professor in the Faculty of Pharmacy, Department of Pharmaceutical Toxicology, Hacettepe University.

Contents

Preface

Botulinum toxin is a neurotoxin produced by the bacterium Clostridium botulinum, an anaerobic, gram-positive, spore-forming rod found on plants and in soil, water, and animal intestinal tracts. It is popularly known as "wonder poison" because it is one of the most toxic biological molecules known to humankind [1].

Botulinum toxin type A has been shown to be useful in the treatment of strabismus in people. Botulinum toxin was later licensed for the treatment of a variety of spastic disorders and other ailments. It is now employed in practically every medical subspecialty. Botox (botulinum toxin-A) was authorized by the FDA in 2002 for the cosmetic purpose of temporarily decreasing glabellar forehead frown lines [1, 2].

The use of botulinum toxin for cosmetic purposes has increased dramatically in recent years, notably since the approval of botulinum toxin-A for the treatment of glabellar lines. Until recently, Botox was mostly used to treat facial expression muscles in the upper one-third of the face [1].

Its current applications include the correction of lines, creases, and wrinkling all over the face, chin, neck, and chest, as well as depressor nasolabial folds, mentalis, medial and lateral brow lifts, to lessen shadows on one's face and maintain a smooth outline of the jaw and cheeks from all directions, and dermatological applications [3]. Botulinum toxins are now used to treat a wide range of medical diseases, including strabismus and focal dystonias, hemifacial spasms, and a variety of spastic movement disorders [4]. In addition, there have been hopeful clinical reports for headaches, hypersalivation, hyperhidrosis, and several chronic illnesses that only respond partially to medical treatment. Botulinum toxin can be used as an alternative to surgery in some cases. In individuals with persistent anal fissures, it appears to be a promising alternative to sphincterotomy and is effective in achalasia. Some autonomic abnormalities that cause gland hypersecretion, such as ptyalism or gustatory sweating, which commonly arise after parotid gland surgery, respond favorably to botulinum toxin [5–9].

Botulinum toxin injections are generally well tolerated, with rare side effects. Idiosyncratic reactions, in general, are uncommon, mild, and temporary. Moderate injection discomfort, local edema, erythema, temporary numbness, headache, malaise, or mild nausea are all possible side effects. Its effect fades as it moves further away from the injection site, but it can extend to surrounding muscles and other tissues. The toxin's activity causes temporary undesired weakness/paralysis of adjacent musculature, which is the most dreaded side effect [1, 10].

The clinical range of type A botulinum toxin has expanded, but the risk of generating antibodies limits the use of high-dose injections on a regular basis. Other botulinum toxin serotypes are being researched as potential replacements. Botulinum toxin type F differs from type A toxin in several ways, including reduced potency, efficacy, and duration of action, as well as blocking a different SNARE protein than type A toxin. To lower the total units and overall antigenic dosage, a combination of toxins A and F has been proposed [1, 11].

Botulinum toxin is a beneficial and effective alternative pain treatment that can be used as a supplement to other treatments or as a standalone treatment. It is a promising possibility for non-opioid pain treatments or for those who do not respond well to opioid treatment since it targets a different receptor type than traditional pain medications like opioids. Botulinum toxin is a generally safe and effective treatment for people suffering from specific types of pain, particularly migraines. More research is needed to better understand the long-term and chronic consequences of botulinum toxin treatment as well as its impact on pregnant, elderly, and adolescent patients.

Dr. Suna Sabuncuoglu
Professor,
Faculty of Pharmacy,
Department of Toxicology,
Hacettepe University,
Sıhhiye, Ankara, Turkey

References

[1] Nigam AN. Dermatol. Botulinum Toxin. 2010;**55**(1):8-14

[2] Münchau A, Bhatia K. Uses of botulinum toxin injection in medicine today. Bmj. 2000;**320**(7228):161-165

[3] Cartee TV, Monheit GD. An overview of botulinum toxins: past, present, and future. Clinics in plastic surgery. 2011;**38**(3):409-426

[4] Dolly JO. Therapeutic and research exploitation of botulinum neurotoxins. European Journal of Neurology. 1997;**4**:S5-S9

[5] Bhatia K, Münchau A, Brown P. Botulinum toxin is a useful treatment in excessive drooling of saliva. Journal of Neurology, Neurosurgery & Psychiatry. 1999;**67**(5):697

[6] Maria G, Cassetta E, Gui D, Brisinda G, Bentivoglio AR, Albanese A. A comparison of botulinum toxin and saline for the treatment of chronic anal fissure. New England Journal of Medicine. 1998;**338**(4):217-220

[7] Relja M, Korsic M, editors. Treatment of tension-type headache by injections of botulinum toxin type A: Double-blind placebo controlled study. Philadelphia, PA, USA: Neurology; 1999

[8] Shelley W, Talanin N, Shelley E. Botulinum toxin therapy for palmar hyperhidrosis. Journal of the American Academy of Dermatology. 1998;**38**(2):227-229

[9] Picelli A, Santamato A, Chemello E, Cinone N, Cisari C, Gandolfi M, et al. Adjuvant treatments associated with botulinum toxin injection for managing spasticity: An overview of the literature. Annals of physical and rehabilitation medicine. 2019;**62**(4):291-296

[10] Bhatia KP, Münchau A, Thompson P, Houser M, Chauhan V, Hutchinson M, et al. Generalised muscular weakness after botulinum toxin injections for dystonia: A report of three cases. Journal of Neurology, Neurosurgery & Psychiatry. 1999;**67**(1):90-93

[11] Mezaki T, Kaji R, Brin MF, Hirota-Katayama M, Kubori T, Shimizu T, et al. Combined use of type A and F botulinum toxins for blepharospasm: A double-blind controlled trial. Movement Disorders: Official Journal of the Movement Disorder Society. 1999;**14**(6):1017-1020

Chapter 1

Medical Applications of Botulinum Toxin: Spasticity

Francisca del Rosario González Núñez,
Francisco Núñez de Castro and Rosa María González Núñez

Abstract

Botulinum toxin is the most powerful toxin ever known. Its huge potential has been used in the last decades to treat some pathologies that attend with acetylcholine's metabolic disturbances. Botulinum toxin effects are not indefinite in time, so it must be applied several times (usually twice a year), thanks to the changes produced by this toxin the rehabilitation of infantile spasticity are easier (the greatest results are seen on abductor muscles, calves, etc.) and on ocular muscles to treat squint. It is also used in some muscle alterations, pathologies of the autonomic nervous system, painful syndromes. The life quality of children with cerebral palsy is improving thanks to this toxin.

Keywords: toxin, acetylcholine, neurotransmitter, applications, therapeutic

1. Introduction

At present, botulinum toxin type A is becoming the treatment of choice for a multitude of pathologies related to alterations in the biochemistry of acetylcholine [1, 2].

Since the middle of the twentieth century, many clinical trials of this type have been carried out. Botulinum toxin A is the most widely used in human therapeutic trials; some of these studies are:

- Therapeutic trials were conducted by A. Scott [3]. In 1973, this author began to use botulinum toxin type A in the treatment of strabismus, initially in non-human primates and since 1980 in humans. He also described its use in endocrine orbital myopathy and lateral rectus palsy.

- Therapeutic trials were conducted by Frueh et al. [4]. They described the use of toxin A in blepharospasm in 1984. In subsequent years toxin injections became the first-line treatment for blepharospasm with very good results (significant improvement in more than 80% of injected patients).

- Therapeutic trials were performed by Tsui et al. [5] and Brin et al. [6]. These investigators reported results of therapeutic trials with toxin A injections for torticollis in patients who had not responded to other treatments and were severely affected.

- Therapeutic trials conducted by Jankovic [7], Gelb [8], and Greene [9] between 1986 and 1991 have conducted at least five blind, placebo-controlled studies focused on toxin A for cervical dystonia. Afterward, they studied it uses in oromandibular, laryngeal, and limb dystonia, confirming its usefulness, particularly in the treatment of closing mandibular dystonia and laryngeal dystonia in adduction.

2. Botulinum toxin

Botulinum toxin represents the most powerful biological toxin known at present.

This toxin is produced by an anaerobic and Gram-positive bacterium, *Clostridium botulinum*, of which 8 immunologically distinct types are known, but only types A, B, and E have been linked to human botulism.

Clostridium botulinum is widely distributed by nature (in soils, mud from lakes or ponds, and in vegetation), so the intestinal contents of fish, birds, and mammals can contain this type of micro-organism. Its spores are quite resistant, especially to heat. To avoid this presence in food, canning industries must use sterilization methods.

Various types of *Clostridium botulinum* are known, each of which produces an immunologically distinct toxin from the others.

These are among the most potent that exist.

One microgram contains 200,000 times the minimum lethal dose for the mice and is approximately the dose lethal for humans.

Types A, B, E, and F are the most common causing human botulism while types C and D cause botulism in poultry and cattle.

Type A is relevant to human pharmacology. It forms a complex with hemagglutinin which can be crystallized.

This is a protein with a molecular weight of 900,000 Daltons. The separation of the hemagglutinin can be carried out without the toxin losing its effectiveness.

It has a neurotoxic fraction made up of protein with a molecular weight of approximately 150,000 Daltons. It is suspected that it is made up of smaller toxic subunits [10].

Botulinum toxin interferes with the conduction of the nerve impulse, once the axonal propagation of the nerve impulse has begun.

The toxin blocks the release of acetylcholine. Therefore, botulinum toxin is an anticholinergic substance which acts as a muscle relaxant and a specific inhibitor of the release of acetylcholine, acting on the presynaptic nerve ending, preventing the action of calcium ions in the exocytosis process necessary for the release of acetylcholine, thus decreasing the plaque potential and causing muscle paralysis [1, 11].

The toxin has two subunits, one of which binds to the membrane receptor responsible for specificity, allowing the entry of the other subunit the one that blocks calcium ions.

In its therapeutic application, due to its form of administration, only neuromuscular transmission interferes at the site of application and the recovery of the nerve impulse takes place gradually as the nerve endings regenerate (**Figure 1**) [10].

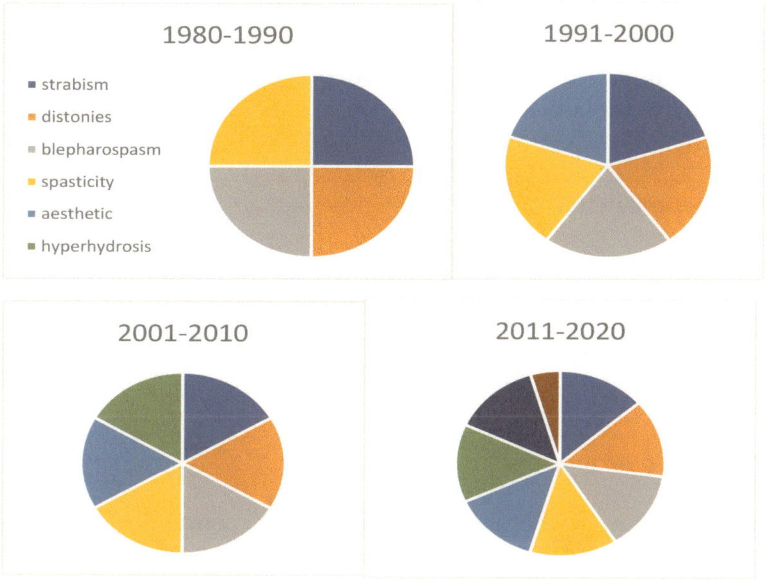

Figure 1.
Principal applications of botulinum toxin in recent decades.

3. Indications

3.1 Applications in muscle disorders

When botulinum toxin is injected into the target tissue, the neurotoxin heavy chain binds to glycoproteins that are specifically found on presynaptic cholinergic nerve receptors. It penetrates the nerve ending of the motor endplate (alpha motor neuron) by endocytosis.

After its internalization, the neurotoxin light chain fragments the SNAP-25 protein, preventing the formation of the SNARE complex (NSF-anchorage protein receptors), which intervenes in the release of acetylcholine by exocytosis.

The blockage of exocytosis becomes definitive on the third day of infiltration and will last until the third month. From the 28th the neuron reacts to the blockage by creating new synaptic buttons (sprouting), restoring the original function. That is, the chemical blockage is irreversible, but the clinical effect is reversible due to the creation of new synaptic buttons [12–16].

Botulinum toxin diffuses locally through the muscle, preferably longitudinally. This local diffusion is dependent on the dose and the precision of the infiltration.

Hence, it is used in pathologies such as:

- Dystonia: Reduced functioning of the neurotransmitter acetylcholine can reduce the excessive functioning of the muscles in different dystonia [17, 18]
    - Cervical dystonia (torticollis)
    - Blepharospasm
    - Apraxia of the eyelid

- Oromandibular-facial lingual dystonia
- Laryngeal dystonia (spasmodic dysphonia)
- Dystonia of the feet, hands
- Occupational cramp
- Facial dystonia
- Meige syndrome
- Tics and stuttering
- Musician dystonia

- Abnormal movements: Botulinum toxin is used to control hyper functional muscular disorders such as [19, 20]
  - Hemifacial spasm
  - Tremor (head, voice, limbs)
  - Myoclonus (palate, spinal)
  - Dystonic motor tics
  - Myoclonus

- Ophthalmological indications: Very good results have been seen when applying botulinum toxin [10]
  - Strabismus (esotropia/exotropia): When the botulinum toxin is injected, a transient paralysis occurs, so that when the extraocular muscle is paralyzed, it relaxes and the antagonist contracts, producing a transient overcorrection of the deviation. This causes an imbalance of forces between antagonists
  - Nystagmus with oscillopsia
  - Ptosis

- Laryngeal disorders [10]
  - Puberphony
  - Granulomas of the vocal processes of the arytenoid cartilage.

- Glandular and secretory cell modulation
  - Growth hormone (acromegaly)

- Inappropriate contractions [18, 20, 21]
    - Spasticity (from stroke or cerebral palsy)
    - Radiculopathy
    - Muscle spasm
    - Stuttering
    - Bruxism
- Smooth muscles [10]
    - Sphincter of oddi dysfunction
    - Chronic spastic bladder
    - Rectosphincteric dyssynergia

3.2 Applications in disorders of the autonomic nervous system

In disorders of the autonomic nervous system, the reduction of the neurotransmitter acetylcholine reduces the functioning of the autonomic nervous system [10, 12, 14] so that the inhibition of acetylcholine release on the postganglionic endings of the sympathetic and parasympathetic systems justifies its clinical use in these pathologies.

- Hyperhidrosis: axillary, palmar, plantar and facial
- Benign prostatic hiperplasia
- Sialorrhea, drooling
- Rhinitis
- Upper esophageal sphincter achalasia/lower esophageal sphincter
- Neurogenic overactive bladder
- Frey's syndrome
- Anal fissures
- Vaginismus/anismus

3.3 Applications in disorders of the afferent nervous systems and pain

The reduction of inflammatory mediators can influence pain syndromes [12, 22–24], due to possible inhibition of the non-selective peripheral release of pain-mediating neurotransmitters (substance P). This justifies its clinical use in these pathologies

- Craniofacial pain
    - Bruxism
    - Temporomandibular pain
    - Trigeminal neuralgia
    - Tension headache and chronic migraine
- Myofascial pain: controversial results
- Painful phantom limb syndrome
- Peripheral ischemia
- Postherpetic neuralgia
- Joint pain: An improvement in these pathologies has been observed when treated with intra-articular injections of botulinum toxin
    - Osteoarthritis
    - Rheumatoid arthritis
- Lumbar pain

4. Spasticity and botulinum toxin

Spasticity consists of "a motor disorder characterized by a speed-dependent increase in the muscle stretch reflex, also called myotatic, with exaggerated pulling on the tendons which is accompanied by hyperreflexia and hypertonia, due to neuronal hyperexcitability being one of the signs of upper motor neuron syndrome, which can be summarized as the abnormal effect of both "tonic and phasic stretch reflexes" [25] (J. Lance, Australian neurologist 1980), which results in an increase in speed-dependent tonic reflexes together with an exaggerated response of tendon (phasic) reflexes in Ref. [25].

This reflex stretching activity can be triggered during voluntary movement when a contraction with shortening of the agonist musculature occurs and is accompanied by a stretching of the antagonist musculature.

4.1 Clinical forms of spasticity

Spasticity presents in characteristic clinical patterns for different neurological etiologies. It should be noted that although the clinical patterns are similar, the response to treatment may vary depending on the etiology.

The most common forms of spasticity are:

4.1.1 Lower limb

- Equine foot, equinovarus
- Digital claw, hyperextension of the first toe
- Thigh adduction
- Hip adducts
- Knee flexion/knee extension

4.1.2 Upper limb

- Shoulder adduction and internal rotation
- Elbow flex
- Wrist flex
- Claw fingers
- Thumb included in the palm

The findings on the examination of spasticity:• Razor resistance (spring)

- Spasticity is directly proportional to stretching speed
- Hyperreflexia with polykinetic response and clonus
- Presence of pyramidal release reflexes and/or spinal automatism in antigravity muscles

If we take into account the age of the patient, we can divide spasticity into two large groups:

A. Infantile spasticity

In childhood, the most common cause of spasticity is infantile cerebral palsy (ICP). An important difference to adult spasticity is that clinical expressiveness in children evolves with growth and causes osteoarticular deformities that interfere with normal development [25].

ICP is classically defined as a persistent movement and posture disorder caused by a non-progressive lesion or defect of the immature brain, before the age of 3–4 years. It has an incidence of 1.5–2.5 per 1000 live births.

However, the current trend is to define ICP by expressing the broad spectrum of its characteristics: ICP is a group of movement and posture disorders that cause activity limitation and is attributed to a non-progressive disorder in fetal development or infant who is frequently accompanied by sensitive, cognition, perception, communication, behavior and/or epilepsy defects.

According to this definition, the classification considers not only the topographic aspect, but also gives special meaning to the severity of the motor impairment, since the functional prognosis will depend not so much on the type of ICP but on the severity.

- Motor disorders:
  - Nature and type (spastic, dyskinetic, ataxic, or mixed)
  - Motor and its severity according to GMFCS (GROSS MOTOR FUNCTION CLASSIFICATION SYSTEM). Gross motor function is a classification system levels for children with cerebral palsy between the ages of 6 and 12 years. It has five levels [26]:
    a. Level I: Walks without restrictions; limitations in more advanced gross motor skills.
    b. Level II: Walks without assistive devices; limitations in walking outdoors and in the community.
    c. Level III: Walks with assistive mobility devices; limitations in walking outdoors and in the community.
    d. Level IV: Self-mobility with limitations; children are transported or use power mobility outdoors and, in the community,
    e. Level V: Self-mobility is severely limited even with the use of assistive technology [26]
- Cause and moment of the disorder.
- Location (hemiplegic, diplegic, or tetraplegic). For the diagnosis of ICP, it is essential to ensure that the causing brain alteration is not progressive.
- That is, to rule out the other degenerative causes of movement disorders. It is recommended to use the guidelines and diagnostic algorithm of the AAN (American Academy of Neurology).

B. Spasticity in adults:

The most common causes of spasticity in adults are [25].

- Acquired brain damage (ABD) caused by head trauma or cerebrovascular accident (CVA)
- Spinal cord injury

The clinical characteristics of spasticity in ABD are:

- Gradual development at 6–8 weeks after stroke and 2–8 weeks after head trauma
- A dysregulation of motor control appears contraction, relaxation
- Pain
- Contraction movements

In addition to the muscle tone disorder, different patterns can be presented:

1. Decortication Pattern: The patient will present shoulder adduction and triple flexion of the upper extremities (elbow, wrist, fingers) and extension of the lower extremities accompanied by equinus.
2. Decerebration Pattern: There is a pattern in the extension of the four extremities with internal rotation of the upper extremities.
3. Mixed Pattern: This is a combination of the two previous patterns.

The appearance of these clinical manifestations has a global impact on the patient with ABD, who will present a decrease or loss of balance and gait, a decrease in manual ability, interference in personal hygiene and activities of daily life (ADL), as well as difficulty in communication and swallowing.

When faced with a patient with ABD, we must consider that, as well as having neuromotor deficits, the patient may also have a set of neuropsychological deficits (cognitive and sensory) therefore, the treatment must be considered in a comprehensive way.

4.2 Spinal cord injury

The causes are multiple: trauma, multiple sclerosis, spinal tumors, infections of vascular origin, familial spastic paraparesis, transverse myelitis, amyotrophic lateral sclerosis and neurofibromatosis.

The incidence of spasticity in spinal cord injuries varies from 65–78% per year of evolution, decreasing to 40%.

Spasticity patterns in spinal cord injury are dependent on the level and degree of injury "classification of the American Spinal Injury Association (ASIA)".

Spasticity in the spinal cord injury is more problematic in incomplete injuries at the cervical level (ASIA grade BCD) and is usually extension spasticity.

5. Evolution of spasticity

Spasticity is not a static phenomenon, it is long-lasting, dynamic, and changing, and there are many factors that influence it, and we must consider it when seeking treatment [27].

Spasticity evolves towards chronicity and is accompanied by static phenomena and alterations in the properties of soft tissues (elasticity, viscosity, plasticity). The alteration of these three properties leads to the establishment of fibrosis of the muscle and of the adjacent structures so that the contracture becomes fixed and retractions and osteoarticular deformities occur, which makes it important to treat as quickly as possible [25, 27].

In the evolution of spasticity, four phases appear, this will determine the treatment.

- SPASTICITY PHASE: In this phase, there is an increase in muscle tone which causes it to be defined as the state of increased tension in a muscle when it is passively lengthened by exaggeration of the muscle stretch reflex.

- VICIOUS ATTITUDE PHASE: This includes muscle imbalance due to the predominance of spasticity in some muscle groups, the most frequent being the plantar and varicose flexors of the foot, the adductors and hip flexors and the elbow, wrist, and finger flexors in the upper limb.

- MUSCLE WITHDRAWAL PHASE: The persistence of these vicious attitudes causes uneven growth between agonist and antagonist muscle groups. This leads to the structuring of this attitude due to the lack of accommodation of the sarcomere, which is unable to achieve normal muscle growth. Understanding by muscle retraction the resistance opposed by the muscle to mobilization when it is not in contraction.

- OSTEOARTICULAR DEFORMATION PHASE: When we are dealing with children in the growth phase, all the pressures and traction stimuli of the growth cartilage will be modified consecutively to the previous phases which, according to Delpech's law, gives rise to osteoarticular deformities (**Figure 2**).

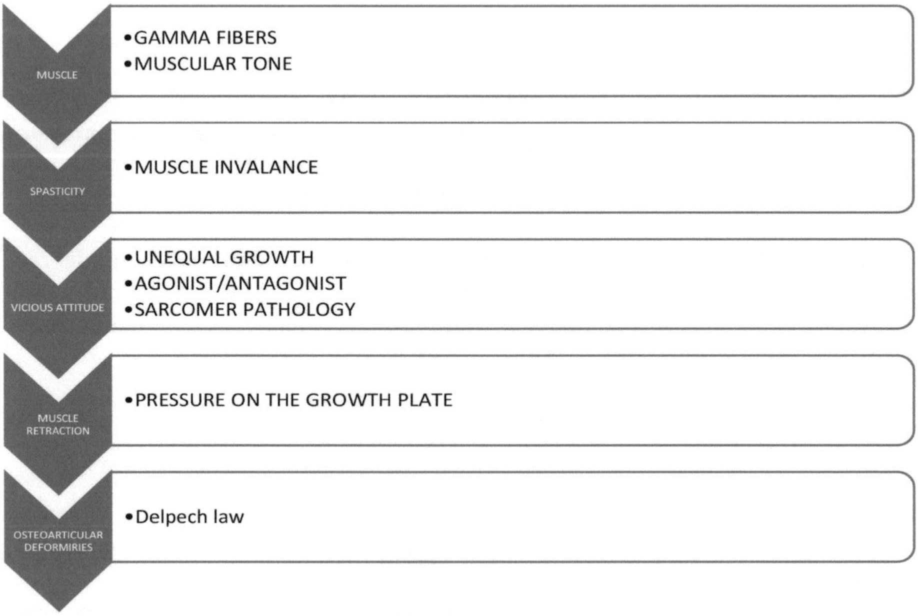

Figure 2.
Delpech law.

It is important to consider that when spasticity occurs in children, it negatively influences their musculoskeletal development, which usually leads to structured deformities, problems in postural control, and limitations to spontaneous mobility [27].

6. Treatment of spasticity

6.1 Objectives in the treatment of spasticity, presentation, indications, application and botulinum toxin dose in spasticity

A well-coordinated, multidisciplinary team will oversee guiding the spasticity treatment, seeking realistic goals that are agreed upon by the patient and the caregiver.

The objectives should be aimed at improving function, reducing pain, preventing complications, and improving hygiene, that is, improving quality of life. In the child, this allows and favors the longitudinal growth of the muscle, avoiding fixed contractures [25–28].

6.1.1 Objectives in the treatment of spasticity

The objectives pursued when applying the toxin on the muscle are:

- Gradually decrease the potential of the drive plate.
- Decrease the state of hyper contraction.
- Muscle relaxation.
- Facilitate extensibility.
- In the case of children, facilitate the longitudinal growth of the muscle.
- An improvement in function, after application in the lower limbs there will be an improvement in gait, greater comfort, balance, and a decrease in falls.

If the application is in the upper limbs there will be greater ease to find it to carry out activities of daily life such as hygiene and food preparation.

- Prevent long-term complications: dislocations, osteoarticular deformities, mainly of the hip, foot, and wrist.
- Improve blood circulation.
- Favors the placement of orthoses and footwear.
- Reduces pain associated with maintained posture.
- An esthetic improvement.
- Facilitate rehabilitation (**Figure 3**).

Figure 3.
Objectives in the treatment with botulinum toxin.

There are different degrees of recommendation for therapeutic strategies for the treatment of spasticity, but we are going to focus on treatment with botulinum toxin.

Botulinum toxin is the treatment of choice for focal spasticity and as a complement in generalized spasticity because it can be administered in the most affected muscles, regardless of its etiology, and is part of the overall treatment.

In the case of generalized spasticity, it should be associated with treatment with intrathecal baclofen or surgery.

As we already know, botulinum toxin acts by blocking the release of acetylcholine at the neuromuscular junction which produces transient chemical denervation, as well as the inhibition of nociceptive neurotransmitters, therefore, playing an analgesic role. The result is transient functional denervation that causes paralysis, muscle atrophy, and electromyographic abnormalities. As already advised, the toxin has two subunits, in such a way that one of them binds to the membrane receptor allowing the second subunit to enter the cell where it will exert a toxic effect by inactivating specific enzymes.

This involves an ADP-riboxylation. The reaction of the toxin is believed to inactivate actin in this way.

It is recognized that only a few molecules are needed to inhibit the release of acetylcholine. The muscle weakness caused by this toxin remains restricted to the injected area, there is histological evidence that toxicity is restricted to the extrafusal muscle fibers, while the intrafusal fibers are relatively free of this affectation.

This causes an alteration in the relationship of the alpha and gamma motor neurons and consequently, there is not only a local paralysis but also an effect on the central motor control mechanisms.

Its effect is progressive, starting in 2 or 3 days, until reaching its maximum per month, maintaining its effect for 3–4 months with a variation interval of 2–6 months.

6.1.2 Presentation and application

Regarding its form of presentation, the only type of botulinum toxin commercially available is type A. In the United States, it is marketed by the Californian laboratory Allergan Pharmaceuticals under the trade name "Botox". It comes in freeze-dried and cold-dried preparations. These are ampoules that must be stored frozen at −5° C so that the toxin is reconstituted at the time of injection with sterile physiological saline (without a preservative). Its potency is expressed in units, in such a way that one unit is equivalent to the amount of toxin capable of killing 50% of a group of female Swiss-Webster mice weighing between 18 and 20 grams (LD50) [21, 26–32].

In the United States, approximately 0.4 nanograms of protein toxin equal 1 unit (or otherwise expressed 2.5 units are equivalent to 1 nanogram). In the case of European botulinum toxin type A, commercially known as "Dysport", the potency of the preparation is different since 1 nanogram is equivalent to 40 units. Due to this divergence in potency in commercial botulinum toxin preparations, the use of doses greater than 500 units per session with the European preparation is not uncommon. The lethal dose of American botulinum toxin type A (Botox) injected into young monkeys is approximately 40 units per kilogram of weight, which, when extrapolated, represents about 50 times the average dose injected for the treatment of focal dystonia. The estimated LD50 in humans is 2500–5000 units according to some authors and closer to 5000 according to others. The diluted solution is collected in a tuberculin syringe and the toxin is injected with a 26–30 gauge needle 0.5 inches long in the superficial muscles and a 22 gauge 1.5 inches long in the muscles. Different dilutions can be prepared depending on the site to be injected, such as 2.5–5 units per 0.1 cc for cervical muscles and 1.25–2.5 units per 0.1 cc for blepharospasm or for a hemifacial spasm. The precise time for its action to appear varies between two and 3 days, reaching its maximum effect 5 or 6 days after the injection, its effect lasting a variable period that extends from 2 weeks to 8 months. (Although we have found some studies that indicate that this duration can be extended to 11 months), this is the time necessary for the toxin binding process, integration of the same, and regeneration of the neuromuscular junction. As already advised, the toxin has two subunits, in such a way that one binds to the membrane receptor allowing the entry of the second subunit into the cell where it will exert a toxic effect by inactivating specific enzymes, for which it involves an ADP reaction—riboxylation.

The toxin is believed to inactivate actin in this way. It is thought that only a few molecules are needed to inhibit the release of acetylcholine.

It is injected into large and superficial muscles, infiltrates the muscle belly by means of anatomical landmarks and palpation. Sometimes it is useful to be guided by electromyography, electrostimulation, or ultrasound. It is important to avoid venous spread.

Two toxins type A (Botox and Dysport) are marketed in Spain. Their doses are not interchangeable.

In children, the dose is calculated in international units per muscle and kilogram of weight. It is also a function of the size of the muscle and the degree of spasticity.

It is diluted in sterile 0.9% physiological saline in 1–2 ml.

The lowest effective dose should be used so that (**Table 1**):

For Botox[r] the dose is 1–6 IU/kg/muscle, with a total dose of up to 12–14 IU/kg, not being able to exceed 300–400 IU.

	Botox	Dysport
Dose IU/kg/muscle	1–6 IU/kg/muscle	3–12 IU/kg/muscle
Total dose	12–14 IU/kg	20–30 IU/kg
Maximum dose	300–400 IU	750–1000 IU

Table 1.
Botulinum toxin dose in spasticity [27].

For Dysport^{r} the doses are 3–12 IU/kg/muscle, with a total dose of up to 20–30 IU/kg, not being able to exceed 750–1000 IU.

The maximum dose with Botox^{r} per injection site is 50 U [21, 26–32].

6.2 Adverse effects

There are some adverse effects produced by this treatment characterized by the appearance of weakness in injected and non-injected muscles or transient bladder paresis (after treatment of hip adductor spasticity), in a few patients, a generalized syndrome with tetraparesis (botulism type), It occurs after application of botulinum toxin and disappears within a period of about 4 weeks [27].

We can also find local adverse effects, which disappear in a few days, such as:

- Injection site pain
- Local inflammatory reaction
- Appearance of flu-like picture
- Diarrhea
- Urinary incontinence
- Allergic reaction

Fatal systemic adverse effects rarely appear, they have been described in isolated cases and with toxin doses that far exceeded the recommended dose.

6.3 Effectiveness

The main infiltration points in the lower limb are triceps, posterior tibial, anterior tibial, abductor of the big toe, abductors, and psoas [1, 21, 31–33].

The main application has been used especially in abductor spasticity of the legs, achieving a reduction in spasticity, pain, and improved hygiene and care of the patient. It has also been used in leg extension spasticity (spastic foot drop) by applying it to the soleus, posterior tibial, and calf muscles, observing an improvement in muscle tone, gait, and foot pain with reduction of the Achilles clonus, in hip dysplasia (caused by the psoas) and of special interest because it can become dislocating. Finally, it has also been shown to be useful in spasticity of the upper extremities [1, 21, 31–33].

DOI: http://dx.doi.org/10.5772/intechopen.101792

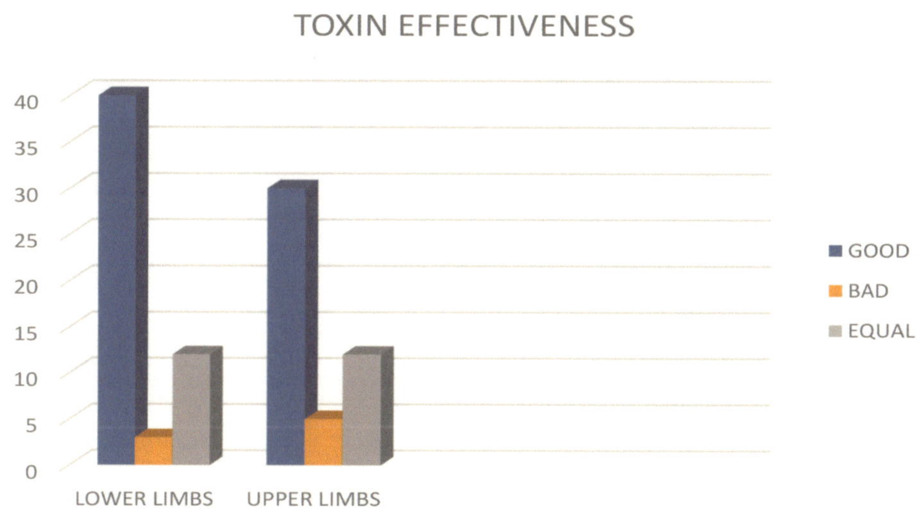

Figure 4.
Effectiveness of botulinum toxin in different limbs.

It should be noted that in general, the results are better in young people, in whom the joints are more dynamic. Although the use of botulinum toxin is very effective for the correction of retractions, as is the case of the Achilles tendon, it should be taken into account before saying its application is not as effective in the upper limbs as in the lower ones, as can be seen in the graph (**Figure 4**).

The most effective points are [1]:

GARCIN'S SCARVED HAND: Wrist flexion with ulnar deviation, thumb adduction, and pronation. The infiltration points are found in the biceps brachii, brachialis anterior, ulnar anterior, palmar, finger flexors (especially superficial ones).

INCLUSION OF THE THUMB: Adductor pollicis on its posterior aspect because the infiltration on its anterior aspect is very painful due to the presence of many sensory receptors.

CHANDELIER PATTERN: The points of infiltration are in the common flexor of the fingers in the pronator quadratus and round, in the posterior ulnar, long supinator (its flexor component), anterior brachialis, and biceps brachii.

TRIPLE FLEXION: Hip flexion: infiltrate the rectus anterior, sartorius, tensor fascia lata, iliac psoas, pectineum, adductors, gluteus minimus and anterior fibers of the gluteus medius. Knee flexion: hamstrings, tensor fascia lata, calves and popliteus. Plantar flexion (equine): triceps, tibialis posterior, long flexor of the fingers.

SCISSOR PATTERN: Hip adduction: pectineal, adductors, internal rectus Internal rotation: adductors.

EQUINE FOOT-VARUS: Posterior tibial, triceps, intrinsic muscles of the foot.

VALGO FOOT: Lateral peroneus brevis and lateral peroneus longus muscles.

7. Contraindications

- Patients allergic to the drug.
- Generalized disorders of muscle function (myasthenia gravis).

- Take blood thinners.
- Inflammation or infection at the injection site.
- Administration of high doses of aminoglycoside antibiotics (especially in patients with renal failure).
- Pregnancy.
- False expectation of cure.
- Uncertainty of a therapeutic follow-up.

Regarding the precautions to be taken for the use of these preparations, the following should be indicated:

- Caution in patients with respiratory disorders during the treatment of Torticollis due to the risk of aspiration.
- During the lactation period, it is not known if it is secreted in significant amounts in breast milk, so as a precaution its use should be avoided in this period [1, 10, 28].

8. Precautions

Caution should be exercised with children with febrile, respiratory, or swallowing processes.

It must be taken into account that its effect is enhanced by aminoglycosides because they intervene in neuromuscular transmission [1, 21, 29].

9. Conclusion

Botulinum toxin offers the advantage that it lacks the systemic side effects of oral drugs such as excessive somnolence or generalized muscle weakness. It allows to offer a local treatment, specifically in the muscle disorder.

Botulinum toxin treatment of spasticity, especially in children, is very effective, always within a global vision. But we cannot forget that it is also a very effective treatment for focal dystonia.

In these cases, its use has been a very effective option, assuming a radical change in the prognosis and in the quality of life of the patients, its use being accepted as routine treatment.

The following table shows a summary of the percentages of efficacy, duration, and complications of the use of this toxin in the different types of focal dystonia:

	Gets better	Beginning	Duration	Complications
Laryngeal adductor	96%	1–2 days	10–16 weeks	Voice dysphagia
Blepharospasm	90%	1–3 days	12–15 weeks	Ptosis

	Gets better	Beginning	Duration	Complications
Cervical	70%	3–7 days	8–12 weeks	dysphagia
Oromandibular	70%	2–7 days	12–14 weeks	dysphagia
Abductive laryngeal	70%	1–2 days	10–16 weeks	voice dysphagia
Hand in hand	60–80%	3–7 days	12–14 weeks	hand weakness

Nowadays it is also not only used for dystonia and spasticity but for upper obstetric brachial palsy (C5-C6), during the first months of life, where it avoids shoulder limitation, and congenital muscular torticollis (infiltrating the sternocleidomastoid and trapezius).

This toxin is used more and more to treat these pathologies improving the life quality of the patients reducing the adverse effects of alternative treatments used in the past.

Author details

Francisca del Rosario González Núñez[1]*, Francisco Núñez de Castro[2] and Rosa María González Núñez[3]

1 Junta de Castilla y Leon, Salamanca, Spain

2 University of Salamanca, Salamanca, Spain

3 Odontolab, Private Dental Clinic, Valladolid, Spain

*Address all correspondence to: kampanilla2002@hotmail.com

References

[1] Gonzalez Nuñez FR. Toxina botulínica y espasticidad. Available from: https://www.efisioterapia.net/articulos/toxina-botulinica-y-espasticidad [Accessed: November 1, 2014]

[2] Erbguth FJ. From poison to remedy: The chequered history of botulinum toxin. Neural trasm (viena). 2008;**115**(4):559-565

[3] Scott AB. Botulinum toxin injection into extraocular muscles as an alternative to strabismus surgery. Journal of Pediatric Ophthalmology and Strabismus. 1980;**17**:21-25. DOI: 10.1016/s0161-6420(80)35127-0

[4] Frueh BR, Felt DP, Wojno TH, Musch DC. Tratamiento del blefaroespasmo con toxina botulínica: Informe preliminar. Archives of Ophthalmology. 1984;**102-110**:1464-1568

[5] Tsui JK, Eisen A, Mac E, Carruthers J, Scott A, Calne DB. A pilot study on the use of botulinum toxin in spasmodic torticollis. The Canadian Journal of Neurological Sciences. 1985;**12**:14-16. DOI: 10.1017/s031716710003540x

[6] Brin MF et al. Localized injections of botulinum toxin for treatment of focal dystonia and hemifacial spasm. Movement Disorders. 1987;**2**(4):237-254. DOI: 10.1017/s031716710003540x

[7] Jankovic J, Orman J. Botulinum A toxin for cranial-cervical dystonia: A double-blind, placebo-controlled study. Neurology. 1987;**37**:616-623. DOI: 10.1212/wnl.37.4.616

[8] Gelb DJ, Lowenstein DH, Aminoff MJ. Controlled trial of botulinum toxin injections in the treatment of spasmodic torticollis. Neurology. 1989;**39**:80-84. DOI: 10.1212/wnl.39.1.80

[9] Greene P, Kaung U, Fahn S, Brin M, Moskowitz C, Flaster E. Double-blind, placebo-controlled trial of botulinum toxin injections for the treatment of spasmodic torticollis. Neurology. 1990;**40**:1213. DOI: 10.1212/wnl.40.8.1213

[10] Lopez del Val LJ, Castro García A. Toxina botulínica: Aplicaciones terapeúticas. ed Masson; 2002. Capítulo 1. pp. 3-23. ISBN 9788445821503

[11] Lenhinger AL, Nelson DL, Cox MM. Principios de bioquímica. 2nd ed. Barcelona: Omega; 1995

[12] Garcia PJ, Ruiz. Aplicaciones de la toxina Botulínica en Neurología. Medicina Clínica (Barcelona). 2013;**141**(1):33-36.2. DOI: 10.1016/j.medcli.2012.12.015

[13] Callejas Pérez MA, Rubio M, Grimalt R, Valls-Solé J. La hiperhidrosis primaria. Medicina Integral. 1996;**36**:223-230. DOI: 10.1016/j.ad.2009.09.004

[14] Jankovic J, Brin MF. Thrapeutic usus of botulin toxin. The New England Journal of Medicine. 1991;**324**:1186-1194. DOI: 10.1056/NEJM199104253241707

[15] Elton JS. Botulinic toxin A in clinical medicine. Journal of Physiology (Paris). 1990;**84**(4):285-289

[16] Hesse S, Lucke D, Malezix M. Botulinum treatment for lower limb espacicity in chronic hemiparetic patients. Journal of Neurology, Neurosurgery, and Psychiatry. 1994;**57**:1.321-1.324

[17] García Ruiz PJ et al. Mitos y evidencias en el empleo de la toxina botulínica: Neurofarmacología y distonías (documento de consenso). Revue Neurologique. 2018;**66**: 5163-5172

[18] Tratamiento con toxina botulínica en la espasticidad, tratamiento con toxina botulínica en los tics. Available from: www.getm.net

[19] Valls-Solé J, Tolosa ES, Martí MJ, Allam N. Treatment with botulinum toxin injections does not change brainstem inter-neural escitability in patients with cervical dystonia. Clinical Neuropharmacology. 1994;7:17-229. DOI: 10.3390/toxins11010021

[20] Infiltración de la toxina botulínica en el espasmo hemifacial. Available from: www.abtusalub.com

[21] Snow BJ, Tsui JKC, Bharr MH, et al. Treatment of spasticity with botulinum toxin: A double-blind study. Annals of Neurology. 1990; **28**:512-515. DOI: 10.1002/ana.41028 0407

[22] Rivera Día RC et al. Toxina botulínica para tratamiento del dolor crónico. Revisión de la evidencia. Revista Colombiana de Anestesiología. 2014;**42**: 205-213. ISSN: 0120-3347

[23] Available from: https//www.ortod oncia.ws/publicaciones/2015/art-34/

[24] Available from: https//www.alcavia. net:ACODEM-informe11

[25] VIllancos-Matellano F et al. Espastidad. Revue Neurologique. 2007; **45**(6):365-375

[26] Downloaded from www.jama.com at McMaster University on February 22, 2009

[27] Cano C. Rehabilitación infantil. Nuevas técnicas de rehabilitación infantil. In: Sociedad Española de rehabilitación infantil 9788498355796. Madrid: Editorial Panamericana; 2013. pp. 122-124. ISBN9788498353440

[28] Aguilar-Barberá M, Calopa-Garriga M, Sánchez-Dalmau B. Indicaciones actuales del tratamiento con tóxina botulínica. Revue Neurologique. 1997;**25**: 138-270

[29] Durif F. Clinical bioequivalence of the current commercial preparations of botulinum toxin. In: En 2° Simposio Europeo sobre toxina botulinica. Cannes; 1995

[30] Carlo RDR, Andrés ALM, Victoria ASM, Sara ES, Manuela GM. Toxina Botulinica para el tratamiento del dolor crónico. Revisión de la evicencia. Revista Colombiana de Anestesiología. 2014;**42**:205-213

[31] González Núñez FR, Núñez Mateos JC, González Núñez RM. Indicaciones médicas de la Tóxina Botulínica. Formación médica continuad. 2021;(2): 118-121

[32] Available from: http://www.servic iofarmaciamanchacentro.es/images/ stories/recursos/recursos/protocolo/ neurologia/2015/protocolo_uso_toxina_ botulinica_17_06_2015.pdf

[33] Análisis de los pacientes con respuesta muy favorable al tratamiento de la espasticidad con toxina botulínica. Available from: http://www.neuropedia tria.org

Chapter 2

The Non-Cosmetic Dermatological Use of Botulinum Neurotoxin

Maria Angelo-Khattar

Abstract

Botulinum neurotoxin injections are currently the most popular non-surgical cosmetic therapy for treating hyperdynamic lines and rebalancing face muscles all over the world. However, there is emerging interest in the use of the neuromodulator for the treatment of non-cosmetic clinical conditions. The present evidence supporting the use of Botulinum toxin in the treatment of acne and oily skin, rosacea, alopecia areata and androgenic alopecia, scar prevention and therapy, postherpetic neuralgia, hyperhidrosis, and disorders exacerbated by sweating is discussed in this chapter. Case reports and a few randomized controlled trials are used to support the use of Botulinum toxin in several of these illnesses. Nonetheless, the therapeutic application of Botulinum toxin in various skin conditions linked with discomfort, inflammation, and hyperhidrosis has a lot of promise.

Keywords: botulinum neurotoxin, rosacea, acne vulgaris, oily skin, hypertrophic scars, keloids, postherpetic neuralgia, idiopathic hyperhidrosis, genodermatoses, psoriasis

1. Introduction

Several serotypes of Botulinum neurotoxin (BoNT) are produced by the bacterium Clostridium botulinum, including BoNT-A, B, C, D, E, F and, G [1]. Currently, only BoNT-A and B are commercially available and, the most widely used isoform in both cosmetic and clinical dermatology is BoNT-A.

To date, only four BoNT-A formulations have received FDA approval for several indications. These include OnabotulinumtoxinA, AbobotulinumtoxinA, IncobotulinumtoxinA and, ProatobotulinumtoxinA [2, 3].

Following the initial approval of BoNT-A for glabellar wrinkles by the US Food and Drug administration in 2002, it has been widely used in cosmetic dermatology, for the treatment of hyperdynamic lines and the rebalancing of facial muscles. BoNT-A is currently the leading non-surgical cosmetic procedure, world -wide [2].

The mechanism of action of importance to the aesthetic use of the neurotoxin is chemical denervation and consequent relaxation of skeletal muscle. Post injection, BoNT-A binds to specific presynaptic receptors on the nerve terminal and subsequently becomes internalised into the presynaptic nerve by receptor-mediated endocytosis. Within the nerve terminal, the light and heavy chain of BoNT-A dissociate by the breakage of the disulfide bonds. The light chain, responsible for the action of the neurotoxin, cleaves SNAP-25 docking protein on the internal neuronal membrane, which plays a key role in the release of acetylcholine during nerve stimulation. Consequently, acetylcholine can no longer be released into the synaptic cleft and muscle stimulation is inhibited until such time as the function of the neuromuscular junction is restored, which normally requires 3 to 6 months [2].

In the past few years, the neuromodulator has gained attention in the field of a clinical dermatology, as an off-label treatment for inflammatory skin diseases and wound healing [4]. The skin is known to interact with the nervous system and there is accumulating evidence that the neurological system has a direct impact on cutaneous inflammation and wound healing. Hence a number of research groups have shown the benefit of BoNT-A injections in acne, rosacea, psoriasis, scar prevention and hypertophic scar treatment, androgenic alopecia and alopecia aerate as well as hyperhidrosis and several conditions exacerbated by sweating [4, 5].

Apart from chemical denervation of cholinergic nerves, other mechanisms have also been postulated as being responsible for the treatment of these conditions includes the inhibition of substance P, glutamate release and calcitonin gene-related peptide (CGRP) and, the suppressions of mast cell activity [6, 7].

This review will address the role of BoNT-A in several non-cosmetic dermatological conditions for which there is current evidence.

2. BoNT-A in oily skin and acne vulgaris

Oily skin is a common dermatological condition reported by many patients, with and without acne. Clinically, the skin appears greasy and unclean with large open pores. Oily skin is troublesome and despite the several topical and systemic treatment options available, it is often difficult to control excessive serum secretion.

The treatment of oily skin with BoNT-A skin was first reported by Shah et al. in 2008 [8]. Twenty subjects with oily skin and large pores were injected intradermally with BoNT-A in the T-zone and evaluated one month post treatment. Photographic evidence of improvement in sebum secretion as well as patient satisfaction was noted in 17 of the 20 subjects [8].

In a prospective study, 25 patients were treated with intradermal injection of BoNT-A on the forehead and followed up for to three months. Sebum production, measured with a sebumeter, revealed a significant reduction in sebum secretion at 1 week and 1, 2 and 3 months after treatment [9]. Furthermore, twenty-one patients (91%) reported that they were satisfied (50–75% improvement) with intradermal botulinum toxin as a treatment for oily skin [9].

A recent split face-controlled study with BoNT-A on one cheek and saline on the other cheek showed a significant decrease in pore size and sebum score at four months in the BoNT-A side as compared to the saline control [10].

Within the sebaceous gland, sebocyte differentiation and sebum secretion are under the control of acetylcholine, since both immature and mature sebocytes express mucarinic acetylcholine receptors. Hence the mechanism of action of BoNT-A is postulated to be due to the inhibition of acetylcholine release into the synaptic cleft, where it normally binds to muscarinic receptors on the postsynaptic membrane [9, 10].

Acne vulgaris, has a prevalence between 82 and 100%, as determined by various studies [11]. Although predominantly believed to be a condition associated with adolescence, it may persist well into adulthood. It is a challenging condition to treat, despite the availability of various topical and systemic options that target the complex pathogenic pathways involved in the development of acne.

The initial step in Acne vulgaris is the formation of the microcomedo [12]. Several factors have been implicated including increased sebum secretion, ductal hypercornification and bacterial colonisation of the pilosebaceous unit by *Cutibacterium acnes* (*C. acnes*).

Hence changes in sebaceous gland function and increased production of sebum are important causative factors in acne [13].

Li et al. showed immunohistochemical and immune-cytofluorescence evidence for the presence of cholinergic receptors in sebaceous glands [14]. Furthermore, the clinical relevance of these findings was assessed by a double-blind, split-face, placebo-controlled study on 20 volunteers. A marked decrease in sebum production was seen in the BoNT-A treated side as compared to the control involunteers with oily skin [14].

Therefore, the modulation of sebum secretion by blocking the activity of acetylcholine in sebocytes and suppression of the sebaceous gland may have a beneficial action in acne, especially in individuals with oily skin [14, 15]. Although there are no systematic clinical trials in this regard, there are some clinical observations to indicate that BoNT-A reduces acne flares. Patients with Tourette syndrome injected with 20–25 units of BoNT-A into the paranasal facial expression muscles, have shown a clearance of perinasal acne one to two weeks post-treatment with improvement persisting for four months [16]. An interesting outcome of a controlled study on BoNT-A treatment for facial wrinkles was the observation that the rate of acne breakouts in the treatment group was lower than that in the placebo group [17].

An additional benefit of BoNT-A in acne may be due to its inhibition of substance P, since it is known that this inflammatory intermediate stimulates lipogenesis in the sebaceous gland and contributes to the onset and aggravation of inflammation in acne [18].

3. BoNT-A in facial flushing and rosacea

Facial flushing, occurs due to the dilatation of facial capillaries which may be idiopathic or secondary to rosacea. It may be sporadic in nature, initiated by agents that act directly on blood vessels as well as due to stimuli such as alcohol, drugs, food additives, and neurological and hormonal stimuli including menopause. In the majority of cases, the treatment of flushing depends on the management of the underlying cause.

Rosacea, on the other hand, is a chronic inflammatory dermatosis affecting approximately 10% of the population and is characterised by persistent erythema affecting the convexities of the skin of the face [19]. It is often accompanied by telangiectasia and exhibits a poor response to the currently available topical, oral and laser treatment options. Rosacea is classified into four major subtypes; Erythemato-telegiectatic, Papulopustular, Phymatous and Ocular rosacea [20]. Erythemto telegiectatic rosacea is associated with persistent erythema and telangiectasia, including episodes of flushing, and is often seen in fair-skinned Caucasian patients. The morphological characteristics of papulopustular rosacea include erythematous papules and pustules on a background of central facial erythema. Phymatous rosacea presents with thickening and hyperplasia of sebaceous glands on the nose. Ocular rosacea, as the name suggests, is associated with blepharo-conjunctivitis and symptoms of dryness, gritty sensation and itching of the eyes [21].

Many rosacea patients have morphological characteristics of more than one subtype of rosacea.

BoNT-A injections have been shown to be of value in the prevention of the erythema and flushing of rosacea. Case studies of two patients treated with intradermal injection of OnabotulinumtoxinA showed improvement in flushing within 2 weeks and the effect lasted for four months post treatment [22].

A case report involving mesotherapy with BoNT-A for the treatment of refractory erythematotelengiectatic and papulopustular rosacea showed significant reduction in erythema, telangiectasia and flushing [23].

In a larger double-blind split-face study, 24 participants with rosacea were randomised to receive intradermal injection of either BoNT-A or normal saline in both cheeks. On the BoNT-A treated side, the Clinician Erythema Assessment (CEA) score significantly decreased, and the Global Aesthetic Improvement Scale (GAIS) score significantly increased up to 12 weeks post treatment [24].

BoNT-A injections may also be used in combination treatment protocols with energy based devices. In a recent study by Al-Niaimi and co-workers, intradermal BoNT-A injections were used following pulsed dye laser treatments for rosacea. Evaluation of erythema index by 3D camera showed a positive synergistic effect of the treatments with a reduction in grading scores of flushing and erythema [25]. In another study, 16 patients were treated with a Tixel device which causes mechanical disruption of the stratum corneum, followed by topical application of 100iu of Abobotulinumtoxin A. significant improvement of the maxameter sore, clinical erythema assessment (CEA) and patient self-assessment was shown for up to 6 months [26].

To date, the pathogenesis of rosacea has not been fully elucidated. However, several mechanisms have been postulated as being involved including acetylcholine release, imbalance of the innate immune system and the skin microbiota as well as abnormal neurovascular signalling [22]. The release of vasoactive neuropeptides including substance P, calcitonin gene-related peptide and vascular endothelial growth factor as well as cathelicidins are believed to lead to rosacea flares. Cathelicidins are antimicrobial peptides that play a major role in the pathogenesis of rosacea [27]. Recently mast cells have been proposed as being responsible for cathelicidin-induced skin inflammation [6].

Hence the postulated mechanism of BoNT-A in the reduction of Rosacea flares could be due to inhibition of acetylcholine release from peripheral nerves in the cutaneous vasodilator system [22], the inhibition of the vasoactive neuropeptides and cathelicidins as well as the inhibition of mast cell degranulation [6, 27].

4. BoNT-A in alopecia

BoNT-A injections have been studied in the treatment of certain non-scarring alopecias including alopecia areata and androgenic alopecia.

Alopecia aerate (AA) is associated with one or more round, smooth bald patches on the scalp. These may be asymptomatic or associated with trichodynia, a burning sensation in the affected area. It may affect females and males at any age and starts in childhood in half of the cases and before 40 years of age in 80% of cases [28]. Alopecia areata is classified as an auto-immune disorder and characterised on histology by bulbar lymphocytic infiltration around terminal hairs [28]. To date the mainstay of treatment of AA is the use of immunosuppressive treatments, corticosteroid injections and minoxidil [29].

Thus far, the only study in the literature on the use of BoNT-A in AA is by Cho et al. [30]. The investigation involved the intradermal injections of 10 U of the neurotoxin in seven patients with AA and the results showed that BoNT-A cannot be used as an alternative treatment for recalcitrant androgenic AA [30]. Nonetheless, since

BoNT-A has an effect on neuro-immunogenic mechanisms, namely the prevention of the release of substance P and calcitonin gene-related peptide, which modulate hair growth [31], further studies concerning the treatment efficacy of BoNT-A for mild to moderate AA are warranted.

Androgenetic alopecia (AGA), is a common type of hair loss in genetic predisposed males and females. It is, in fact, the major causes of progressive hair thinning and affects 50% of females and 80% of the male population [32].

The current hypothesis for the pathophysiology of androgenetic alopecia involves the role of dihydrotestosterone (DHT), a metabolite of testosterone, as the causative factor. DHT is synthesised from testosterone via type II 5-alpha reductase enzyme (5-αR2). DHT accumulates in AGA-prone hair follicles which are sensitive to DHT resulting in the miniaturisation of the hair follicles and ultimately hair thinning, finally culminating in AGA. Consequently, most treatments for AGA target 5 alpha reductase enzyme. The commonly prescribed effective therapy for AGA is Finasteride. The drug reduces serum DHT and scalp tissue DHT by inhibiting 5-αR2. Apart from the genetic predisposition, hypoxia of the scalp is believed to increase the concentrations of DHT by reducing its clearance from the blood [33–35].

Pilot studies [33–35] have shown injections of BoNT-A in various areas of the scalp to be effective in the treatment of AA, resulting in an increase in mean hair count and a reduction in hair loss. The mechanism responsible has been postulated to be due to the relaxation of scalp muscles, thereby relieving scalp tension and resulting in a vasodilator effect. Ultimately the improved perfusion of the scalp along with the enhanced clearance of DHT are responsible for the potential efficacy in the treatment of AGA. A further benefit of the enhanced perfusion is the increased conversion of testosterone to oestradiol, which is known to enhance hair growth.

However, despite the promising results shown in these studies, further investigations such as randomised controlled trials are necessary to establish the benefit of BoNT-A in AGA.

5. Hypertophic scars and keloids

Scars and keloids are a consequence of a dysregulation of the normal wound healing reaction. Wound healing occurs through organised sequential steps of haemostasis, inflammation, proliferation and tissue remodelling. Each of steps of the wound healing cascade is dependent upon several cytokines and growth factors. Interaction among these processes results in the synthesis of collagen over the surface of the wound. On a contrary note, an alteration in the molecular factors, complex network of pro-fibrotic, and anti-fibrotic molecules such as growth factors, proteolytic enzymes, and extracellular matrix proteins may lead to the formation of hypertrophic scars and keloids [36].

Scars have a negative effect on the patient's quality of life as they may impact on the patient both functionally and cosmetically. There are currently various therapeutic approaches in the management of scars including silicone dressings, energy-based devices, surgical excision, cryotherapy and intralesional steroid injections. However, to date, there is not definitive cure and typically combination approaches are employed for the management of scars [37].

The mechanism of action of BoNT-A in the prevention of hypertrophic and keloid scar formation is partially believed to be due to a reduction in fibroblast proliferation and the expression of the inflammatory cytokineTGF-β, one of the main intermediates responsible for scar formation [38]. The neuromodulator also suppresses the proliferation of fibroblasts in hypertrophic scar tissue [39].

An additional contributing factor to hypertrophic scar and keloid formation is tension on the scar [40]. Hence, BoNT-A may modulate and possibly prevent scar formation by reducing muscle contraction and decreasing skin tension in the scar area.

In a double-blind, split-scar randomised controlled trial in 15 patients with post-thyroidectomy scars. Patients were injected within 10 days of surgery and evaluated at 6 months. Results showed a significant improvement in the BoNT-A treated half of the scar as compared to the saline-treated control [41].

A recent randomised controlled trial in 24 patients with keloids revealed that 5 iu/sq.cm of BoNT-A, injected every 8 weeks for a total of three session, was as effective in softening the keloids and reducing their volume, as intralesional corticosteroids, which are considered the first-line treatment for keloids [42]. Both lesion height and volume were decreased in both cases, however BoNT-A was found to be superior in reducing pain and itch associated with keloids [42]. However, in a prospective uncontrolled study by Gauglitz, patients were injected every 2 months for up to 6 months and the results showed no differences in expression of markers and no regression of keloid tissue was found [43].

Despite the inconsistency in results found in the various studies to date, the role of BoNT-A in the prevention and treatment of hypertrophic scars and keloids does show potential. As such, more structured large- scale double-blind, controlled trials and long-term follow up are warranted to determine the value of BoNT-A in the management and prevention of hypertrophic scars and keloids.

6. Postherpetic neuralgia

Postherpetic neuralgia (PHN) is a very painful conditions which typically occurs post herpes zoster virus infection and can last up to three months, following the initial viral infection. The local neuropathic pain and the sleep disturbances associated with PHN, impacts negatively on the patients quality of life. Conventional treatment approaches include topical agents such as capsaicin and topical anaesthetics as well as nonsteroidal anti-inflammatory drugs, gabapentin and tricyclic antidepressants [44]. However the pain and discomfort may be resistant to all of these drugs.

Xiao et all enrolled 60 patients in a randomised, double-blind, placebo-controlled study whereby patients were divided into three groups; BoNT-A group, saline group and 0.5% lidocaine group. All patients were treated once, with the BoNT-A group receiving a maximum dose of 200iu by intra and subcutaneous injections. The patients were followed up for a period of three months. Of the three groups, the BoNT-A group exhibited the most significant improvement in the Visual Analog Scale (VAS) and sleep quality [45].

In another randomised control trial where 30 patients with PHN were treated with either placebo or a total dose of 200iu BoNT-A, the BoNT-A patients showed a significant reduction in Visual Analog Scale pain sores and which lasted for 16 weeks [46].

The mechanism of pain reduction by BoNT-A is believed to be due to the inhibition of various inflammatory intermediates including calcitonin gene-related peptide, glutamate and substance P.

7. Idiopathic Hyperhydrosis and conditions exacerbated by sweat production

BoNT-A is an established treatment option for primary axillary and focal idiopathic hyperhydrosis and in 2004 OnabotulinumtoxinA received FDA approval for severe primary axillary hyperhidrosis [47, 48]. BoNT-A inhibits the release of acetylcholine from sympathetic nerve fibres that stimulate the eccrine sweat glands and leads to a decrease in sweat production. Several dermatological conditions are exacerbated by sweat and these include Hidradenitis suppurativa, pompholyx, and several genodermatoses including Haily-Haily disease, Darrier disease, Epidermolysis bullosa simplex and Epidemolysis bullosa simples Weber-Cockayne Type, and Pachyonychia Congenita.

Hidradenitis suppurativa (HS) also knows as acne inversa, often starts at puberty and is most active between the ages of 20 and 40 years. It is a a chronic inflammatory condition that affects the apocrine glands-bearing areas in the groin, axillae, and under the breasts. It presents clinically as persistent and recurring painful papules, nodules, abscesses and sinus tracts. The condition is associated with a purulent discharge and culminates in both hypertrophic and atrophic scarring. The pain and psychosocial impact of the disease may be debilitating to HS patients, and many suffer from depression and a negative body image [49].

The first case report of HS of the axillae treated with BoNT-A was in 2005 in a young female who was successfully treated with a single dose of the neurotoxin and remained symptom free for ten months [50].

Thereafter, other case reports confirmed the efficacy of BoNT-A in cases of recalcitrant HS [51–53].

The precise mechanism of action of BoNT-A in HS is unclear but it is believed to be due to the inhibition of sweat production. A moist environment in the inguinal folds and axillae predisposes to maceration and the proliferation of bacteria that precipitate the inflammation and symptoms of HS. Hence the maintenance of a dry environment in the affected areas reduces the population of skin flora and potentially relieves the condition [54].

Pompholyx also known as dyshidrotic eczema is a bullous disease of the palms and soles. The condition is associated with pain, pruritis and discomfort, especially when wearing gloves and shoes. Patients are prone to bacterial and fungal infections and sweat and occlusion are predisposing factors to the condition [55].

Improvement in vesiculation and itch has been shown with intradermal injections of BoNT-A in seven out of ten patients, where one hand was injected with 100 iu of the neurotoxin and the other hand was injected with a saline control. The efficacy of BoNT-A in pompholyx is not solely due to its anhidrotic action but due to its inhibition of substance P [55].

8. Genodermatoses

Certain genodermatoses such as *Hailey-hailey disease (HH), Dariers disease, Epidermolysis Bullosa Simplex(EBS)*, and *Pachyonychia Congenita* (PC) are, in the main, exceedingly painful conditions which have no curative treatments, despite several therapeutic options such as systemic and interventional therapies. These conditions are exacerbated by sweating. Hence BoNT-A has been successfully used for this purpose in several of these disorders. Apart from sweat reduction, the known modulation of the neurotoxin on pain-related neurotransmitters, contributes to the pain relief and decrease in lesions in these conditions.

Hailey-Hailey (HH) disease also known as Familial Benign Pemphigus is characterised by flaccid blistering lesions and erosions in the intertriginous parts of the body. This condition is exacerbated by sweat and bacterial colonisation, hence the reduction of sweat by BoNT-A has been proven to be of value in the attenuation of the disease. Several case reports using various regimens have shown marked improvement in all cases after injections of BoNT-A [56–58].

Recently a retrospective study of eight patients who were treated with topical BoNT-A post application of the Tixel thermo-mechanical ablative system, showed positive results, with seven patients (87.5%) experiencing good or partial response [59].

A recent systematic review of sixteen publications including a total of 38 patients revealed that of all the cases, only one patient did not have a response while all the other patients had partial or complete remission.

BoNT-A was found to be an encouraging treatment option in recalcitrant HH disease [60].

Darier disease is an autosomal dominant genetic disorder characterised by warty papules in seborrheic and flexural areas. The signs and symptoms of Darier disease vary significantly between individuals. Some patients display minor signs that are asymptomatic whilst others have widespread lesions which may be extremely painful, associated with itch and a distressing malodour.

Case reports on the use of BoNT-A injections have been shown to significantly improve the lesions and symptoms of Dariers disease.

A recent open label 6-month interventional study in three patients with the neurotoxin resulted in a reduction in skin lesions in the affected area and consequently had a positive impact on the patients quality of life [58].

Epidermolysis bullosa simplex(ES) and Epidermolysis bullosa simplex-Weber Cockrayne (EBS-WC) are conditions associated with keratinocyte fragility resulting in bullae, hyperkeratosis and plantar pain. The conditions are exacerbated by warmth and sweat. Several cases of EBS and EBS-WC have been reported in the literature that were successfully managed by BoNT-A. The reduction in pain, blister formation and callosities were found on average, to persist for 3 months [61–63].

Pachionychia congenita (PC) is a rare genetic dermatosis that is characterised by hypertrophic nails and plantar hyperkeratosis. In certain cases, the condition causes severe neuropathic pain, to the extent that some patients are unable to walk and require mobility assistance, such as a wheelchair. The condition is exacerbated by heat and sweat.

In a case series, three patients with PC injected with BoNT-A, experienced decrease in pain and discomfort [64]. A report of two cases treated with plantar injection of BoNT-A for PC, resulted in marked improvement for a duration of six months with no adverse effects or tachyphylaxis [65].

In a recent study, Koren et al. treated five patients with PC and found significant improvement of between 20 and 72% in the PC specific quality of life questionnaire. The duration of effect lasted between 3 and 4 months [66].

9. Psoriasis

Plaque Psoriasis is an inflammatory disease that presents as small to large, well-demarcated dry and red scaly plaques covered with silvery flakes, which may be painful and itchy. Typically, the lower back, trunk, elbows and knees are involved but it may occur on any part of the body. Inverse psoriasis, sometimes also known as intertriginous psoriasis, is a form of psoriasis that affects skin folds.

Psoriasis is often associated with nail and scalp psoriasis. The condition is associated with multiple comorbidities and markedly diminishes the patients' quality of life.

Whilst topical therapies remain the basis for treating mild psoriasis, currently, biologics that inhibit inflammatory intermediates such as TNF-α and IL-17 are the treatments of choice for moderate to severe disease [67].

Psoriasis is associated with high levels of Calcitonin Gene-related peptide and Substance P and consequently BoNT-A has been shown to be effective in managing both, consequently reducing both itch and pain [68, 69].

G Gonzales et al. treated eight patients with stable and recalcitrant plaque psoriasis, with subcutaneous injections of AbobotulinumtoxinA at a dose of 5 units every square cm. The patients were then evaluated at 2 and 4 weeks after treatment. The outcome was a significant improvement in all patients, 4 weeks after treatment, with no significant side effects [70].

In a split body comparative study, thirty five patients with chronic plaque psoriasis were treated with either intradermal BoNT-A or 5-fluorouracil (5-FU). Treatment outcomes were assessed by two blinded observers and the criteria included induction scores, erythema, scaling and induration were evaluated. No significant difference was found between the two sides regarding the clinical response. The response rate was found to be 85% on the BoNT-A side and 90% on the 5-FU side [71].

10. Conclusion

In this chapter the current off-label use of BoNT-A in various clinical, non-cosmetic dermatological conditions, has been reviewed. The neuromodulator has a complex mechanism of action, which to date has not been fully elucidated. However, it is clear that there is great potential for its therapeutic use in various skin condition associated with pain, inflammation and with hyperhydrosis.

Many of these disorders are chronic conditions and consequently will require repeat injections of BoNT-A, hence it is important to establish the long-term safety and efficacy of the neuromodulator.

The long-term safety of BoNT-A is well-established, yet there remains the potential for tachyphylaxis with cyclical treatments since we know that BoNT-A has an immunogenic potential.

Experience with BoNT-A for cosmetic indications, where doses of up to 100 units are injected every 4–6 months, has shown a very low incidence of immunoresistance. Hence we may assume that the development of neutralising antibodies to BoNT-A is very rare. However, it appears that the key factors in this regard may be the treatment frequency and the dose injected per session.

The evidence to date for the clinical efficacy of BoNT-A in many of these conditions is based on Level V case reports, with diverse treatment protocols, and a very few randomised controlled trials.

Hence further systematic double-blind randomised trials with larger patient populations are warranted to establish the role of BoNT-A in these clinical non-cosmetic dermatological conditions. Furthermore, basic research into the role of BoNT-A in the modulation of neuropeptides and hence it effects on the cutaneous neuroimmune system are required. In conclusion, a consensus on the injection protocol and dosing regimen for each indication is essential.

Author details

Maria Angelo-Khattar[1,2]

1 Altaderma Clinic, Dubai, UAE

2 American Academy of Anti-Aging Medicine, Dubai

*Address all correspondence to: mkhattar@younatagroup.com

References

[1] Von Berg L et al. Functional detection of botulinum neurotoxin serotypes a to F by monoclonal neoepitope-specific antibodies and suspension array technology. Scientific Reports. 2019;**9**:5531

[2] Schlessinger et al. New uses of abobotulinumtoxinA in aesthetics. Aesthetic Surgery Journal. 2017;**37**:545-558

[3] Hilton L. Jeuveau takes aim at aesthetic market share. Accessed: April 28, 2020. Available from: https://www.dermatologytimes.com/article/jeuveau-takes-aim-aesthetic-market-share

[4] Guida S et al. New trends in botulinum toxin use in Dermatology. Dermatology Practical and Conceptual. 2018;**8**:277-282

[5] Campanati A et al. Botulinum toxin off-label use in dermatology: A review. Skin Appendage Disorders. 2017;**3**:39-56. DOI: 10.1159/000452341

[6] Choi JE, Werbel T, Wang Z, Wu CC, Yaksh TL, Di Nardo A. Botulinum toxin blocks mast cells and prevents rosacea like inflammation. Journal of Dermatological Science. 2019;**93**(1):58-64. DOI: 10.1016/j. jdermsci.2018.12.004

[7] Carmichael MM, Dostrovsky JO, Charlton MP. Peptide-mediated transdermal delivery of botulinum neurotoxin type a reduces neurogenic inflammation in the skin. Pain. 2010;**149**:316-324

[8] Shah AR. Use of intradermal botulinum toxin to reduce sebum production and facial pore size. Journal of Drugs in Dermatology. 2008;**7**(9):847-850

[9] Rose AE, Goldberg DJ. Safety and efficacty of intradermal injection of botulinum toxin for the treatment of oily skin. Dermatologic Surgery. 2013;**39**:443-448

[10] Sayed K et al. The efficacy of intradermal injections of botulinum toxin in management of enlarged facial pores and seborrhea: A split face controlled study. The Journal of Dermatological Treatment. 2019;**32**:1-23

[11] Dreno B. Recent data on epidemiology of acne. Annales de Dermatologie et de Vénéréologie. 2010;**137**:549-551

[12] O'Toole EA, Mellerio JE. Wound healing. In: Burns T, Breathnack S, Cox N, Griffiths D, editors. Rooks Textbook of Dermatology. 8th ed. Oxford: Wiley-Blackwell; 2010. pp. 14:1-14.27

[13] Shamlou G. Khachmoune a an update review of the sebaceous gland and its role in health and disease Part1: Embryology, evolution, structure and function of sebaceous glands. Dermatologic Therapy. 2021;**34**(2):e14862

[14] Li ZJ, Park SB, Sohn KC, et al. Regulation of lipid pro- duction by acetylcholine signalling in human sebaceous glands. Journal of Dermatological Science. 2013;**72**(2): 116-122

[15] Hazarika N. Acne vulgaris: New evidence in pathogenesis and future modalities of treatment. Journal of Dermatological Treatment. 2021;**32**:277-281

[16] Diamond A, Jankovic J. Botulinum toxin in dermatology—Beyond wrinkles and sweat. Journal of Cosmetic Dermatology. 2006;**5**:169

[17] Brin MF, Boodhoo TI, Pogoda JM, James LM, Demos G, Terashima Y,

et al. Safety and tolerability of onabotulinumtoxinA in the treatment of facial lines: A meta-analysis of individual patient data from global clinical registration studies in 1678 participants. Journal of the American Academy of Dermatology. 2009;**61**:961-970

[18] Toyoda M, Morohashi M. New aspects in acne inflammation. Dermatology. 2003;**206**:17-23

[19] Tan J, Schofer H, Araviiskaia E, Audibert F, Kerrouche N, Berg M. RISE study group. Prevalence of rosacea in the general population of Germany and Russia – The RISE study. Journal of the European Academy of Dermatology and Venereology. 2016;**30**:428-434

[20] Wilkin J, Dahl M, Detmar M, Drake L, Feinstein A, Odom R, et al. Standard classification of rosacea: Report of the national rosacea society expert com- mittee on the classification and staging of rosacea. Journal of the American Academy of Dermatology. 2002;**46**:584-587

[21] Zhang H et al. Rosacea treatment: Review and update. Dermatologic Therapy. 2021;**11**:13-24

[22] Dayan SH, Pritzker RN, Arkins JP. A new treatment regimen for rosacea: onabotulinumtoxinA. Journal of Drugs in Dermatology. 2012;**11**(12):e76-e79

[23] Bharti J, Sonthalia S, Jakhar D. Mesotherapy with botulinum toxin for the treatment of refractory vascular and papulopustular rosacea. Journal of the American Academy of Dermatology. 2018;**19**:S0190-S9662

[24] Kim MJ, Kim JH, Cheon HI, et al. Assessment of skin physiology change and safety after intradermal injections with botulinum toxin: A randomized, double-blind, placebo-controlled, split-face pilot study in rosacea patients with facial erythema. Dermatologic Surgery. 2019;**45**:1155-1162

[25] Al-Niaimi F, Glagoleva E, Araviiskaia E. Pulsed dye laser followed by intradermal botulinum toxin type-a in the treatment of rosacea-associated erythema and flushing. Dermatologic Therapy. 2020;**33**:e13976

[26] Friedman O et al. The toxic edge- a novel treatment for refractory erythema and Flushing of rosacea. Lasers in Surgery and Medicine. 2019;**50**:325-331

[27] Ahn CS, Huang WW. Rosacea pathogenesis. Dermatologic Clinics. 2018;**36**(2):81-86

[28] Pratt et al. Alopecia areata. Nature Reviews. 2017;**3**:17011

[29] Strazzulla LC, Wang EHC, Avila L, et al. Alopecia areata: An appraisal of new treatment approaches and overview of current therapies. Journal of the American Academy of Dermatology. 2018;**78**:15-24

[30] Cho HR, Lew BL, Lew H, Sim WY. Treatment effects of intradermal botulinum toxin type a injection on alopecia areata. Dermatologic Surgery. 2010;**36**:2175-2181

[31] Paus R, Heinzelmann T, Schultz KD, Furkert J, Fechner K, Czarnetzki BM. Hair growth induction by substance P. Laboratory Investigation. 1994;**71**:134-140

[32] English RS Jr. A hypothetical pathogenesis model for androgenic alopecia: Clarifying the dihyrotesterone paradox and rate-limiting recovery factors. Medical Hypotheses. 2018;**111**:73-81

[33] Freund BJ, Schwartz M. Treatment of male pattern bald- ness with botulinum toxin: A pilot study.

Plastic and Reconstructive Surgery. 2010;**126**:246e-248e

[34] Singh S, Neema S, Vasudevan B. A pilot study to evaluate Ef- fectiveness of botulinum toxin in treatment of androgenetic alopecia in males. Journal of Cutaneous and Aesthetic Surgery. 2017;**10**:163-167

[35] Zhou Y, Yu S, Zhao J, et al. Effectiveness and safety of botulinum toxin type a in the treatment of androgenetic alopecia. BioMed Research International. 2020;**2020**:1501903

[36] Limandjaja G, Neissen FB, Scheper RJ, Gibbs S. Hypertrophic scars and keloids: Overview of the evidence and practical guide for the differentiation between these abnormal scars. Experimental Dermatology. 2021;**30**:146-161

[37] Campanati A, Martina E, Giuliodori K, Consales V, Bobyr I, Offidani A. Botulinum toxin off-label use in dermatology: A review. Skin Appendage Disorders. 2017;**3**:39-56

[38] Forbat E, Ali FR, Al-Niaimi F. Non-cosmetic dermatological uses of botulinum neurotoxin. Journal of the European Academy of Dermatology and Venereology. 2016;**30**:2023-2029

[39] Xiao Z, Zhang M, Liu Y, Ren L. Botulinum toxin type a inhibits connective tissue growth factor expression in fibroblasts derived from hypertrophic scar. Aesthetic Plastic Surgery. 2011;**35**:802-807

[40] Elhefnawy AM. Assessment of intralesional injection of botulinum toxin type a injection for hypertrophic scars. Indian Journal of Dermatology, Venereology and Leprology. 2016;**82**:279-283

[41] Kim YS, Lee HJ, Cho SH, Lee JD, Kim HS. Early postoperative treatment of thyroidectomy scars using botulinum toxin: A split-scar, double-blind randomized controlled trial. Wound Repair and Regeneration. 2014;**22**:605-612

[42] Shaarawy E, Hegazy RA, Abdel Hay RM. Intralesional botulinum toxin type a equally effective and better tolerated than intralesional steroid in the treatment of keloids: A randomized controlled trial. Journal of Cosmetic Dermatology. 2015;**14**:161-166

[43] Gauglitz GG, Bureik D, Dombrowski Y, Pavicic T, Ruzicka T, Schauber J. Botulinum toxin a for the treatment of keloids. Skin Pharmacology and Physiology. 2012;**25**:313-318

[44] Ri S, Kivi A, Wissel J. The safety and effect of local Botulinumtoxin a injections for long-term Management of Chronic Pain in post-herpetic neuralgia: Literature review and cases report treated with Incobotulinumtoxin a. Journal of Personalized Medicine. **202**(11):758

[45] Xiao L, Mackey S, Hui H, Xong D, Zhang Q, Zhang D. Subcutaneous injection of botulinum toxin a is beneficial in postherpetic neuralgia. Pain Medicine. 2010;**11**:1827-1833

[46] Apalla Z, Sotiriou E, Lallas A, Lazaridou E, Ioannides D. Botulinum toxin a in postherpetic neuralgia: A parallel, randomized, double-blind, single-dose, placebo-controlled trial. The Clinical Journal of Pain. 2013;**29**:857-864

[47] NaumannM LNJ. Botulinum toxin type a in treatment of bilateral primary axillary hyperhidrosis: Randomised, parallel group, double blind, placebo controlled trial. BMJ. 2001;**323**:596

[48] Schnider P, Moraru E, Kettler H, Binder M, et al. Treatment of focal

hyperhidrosis with botulinum toxin type a: Long term follow up in 61 patients. British Journal of Dermatology. 2001;**145**:289-293

[49] Buimer MG, Woobes T, JHG K. Hidradenitis Sup. British Journal of Surgery. 2009;**96**:350-360

[50] O'Reilly DJ, Pleat JM, Richards AM. Treat- ment of hidradenitis suppurativa with botuli- num toxin a. Plastic and Reconstructive Surgery. 2005;**116**:1575-1576

[51] Feito-Rodríguez M, Sendagorta-CudÓs E, Herranz-Pinto P, De Lucas-Laguna R. Prepu- bertal hidradenitis suppurativa successfully treated with botulinum toxin a. Dermatologic Surgery. 2009;**35**: 1300-1302

[52] Khoo ABS, Burova EP. Hidradenitis suppura- tiva treated with *clostridium botulinum* toxin a. Clinical and Experimental Dermatology. 2014;**39**:749-750

[53] Martina E, Offidani A. Hidradenitis suppurativa: How to treat with BoNT-A. In: Campanati A, Offidani A, editors. Botulinum Toxin in Dermatology. Hauppauge: Nova Science Publishers; 2015. pp. 61-75

[54] Hua VJ, Kuo KY, Cho HG, Sarin KY. Hyperhidrosis affects quality of life in hidradenitis suppurativa: A prospective analysis. Journal of the American Academy of Dermatology. 2020;**82**:753-754

[55] Swartling C, Naver H, Lindberg M, Anveden I. Treatment of dyshidrotic hand dermatitis with intradermal botulinum toxin. Journal of the American Academy of Dermatology. 2002;**47**:667-671

[56] Kang NG, Yoon TJ, Kim TH. Botulinum toxin type a as an effective adjuvant therapy for Hailey-Hailey disease. Dermatologic Surgery. 2002;**28**:543

[57] Bessa GR, Grazziotin TC, Manzoni AP, Weber MB, Bonamigo RR. Hailey-Hailey disease treatment with botulinum toxin type a. Anais Brasileiros de Dermatologia. 2010;**85**:717-722

[58] Dreyfus I, Maza A, Rodrigues L, Merlos M, Texier H, Rousseau V, et al. Botulinum toxin injections as an effective treatment for patients with intertriginous Hailey-Hailey or Darier disease: An open-label 6-month pilot interventional study. Orphanet Journal of Rare Diseases. 2021;**16**:93

[59] Bar-Ilan E, Koren A, Shehadeh W, Mashiah J, Sprecher E. Artzi O an enhanced transcutaneous delivery of botulinum toxin for the treatment of Hailey-Hailey disease. Dermatologic Therapy. 2020;**33**:e13184

[60] Zhang H, Tang K, Wang Y, Fang R, Sun W. Botulinum toxin in treating Hailey-Hailey disease: A systematic review. Journal of Cosmetic Dermatology. 2021;**20**:1396-1402

[61] Holahan HM et al. Treatment of symptomatic epidermolysis bullosa simplex with botulinum toxin in a pediatric patient. JAAD Case Reports. 2016;**2**:259-260

[62] Abitbol R, Zhou L. Treatment of epidermolysis bullosa simples, weber cockayne type, with botulinum toxin type a. Archives of Dermatology. 2009;**45**:13-15

[63] Swartling C et al. Botulinum toxin in the treatment of sweat-worsening foot problems in patients with epidermolysis bullosa simplex and pachyonychia congenita. The British Journal of Dermatology. 2010;**163**:1072-1076

[64] Swartling C, Vahlquist A. Treatment of pachyonychia congenita with plantar injec- tions of botulinum toxin. The British Journal of Dermatology. 2006;**154**:763-765

[65] Gonzalez-Ramos J et al. Efficacy of botulinum toxin in pachyonychia congenita type I : Report of two new cases. Dermatologic Therapy. 2016;**29**:32

[66] Koren A, Sprecher E, Reider E, Artzi O. A treatment protocol for botulinum toxin injections in the treatment of pachyonychia congenita associated keratoderma. BMJ. 2020;**182**: 671-677

[67] Arnstrong AW, Read C. Pathophysiology, clinical presentation, and treatment of psoriasis: A review. JAMA. 2020;**323**:1945-1960

[68] Ward NL, Kavlick KD, Diaconu D, Dawes SM, Michaels KA, Gilbert E. Botulinum neu- rotoxin a decreases infiltrating cutaneous lymphocytes and improves acanthosis in the KC-Tie2 mouse model. The Journal of Investigative Dermatology. 2012;**132**:1927-1930

[69] Zanchi M, Favot F, Bizzarini M, Piai M. Do- nini M, Sedona P: Botulinum toxin type-a for the treatment of inverse psoriasis. The Journal of the European Academy of Dermatology and Venereology. 2008;**22**:431-436

[70] Gonzalez D, Franco M, Londono A, Valenzuela F. Breaking paradigms in the treatment of psoriasis: Use of botulinum toxin for the treatment of plaque psoriasis. Dermatologic Therapy. 2020;**33**:e14319

[71] Khattab FM, Samir MA. Botulinum toxin type-a versus 5-fluorouracil in the treatment of plaque psoriasis: Comparative study. Journal of Cosmetic Dermatology. 2021;**20**:3128-3132

Chapter 3

The Pharmacology of Botulinum Toxin Type A

Anna V. Reznik

Abstract

The aim of this chapter is to structure current information clarifying the most disputable issues of botulinum neurotoxin type A (BoNT/A) pharmacology after systemic (botulism) impact and local medical application. Botulinum neurotoxin (BoNT) pharmacological features evaluated open ways to study factors affecting its biological activity: to extend/shorten its effect duration, to increase/decrease BoNT sensitivity in specific patient populations. The chapter presents unique molecular mechanisms underlying BoNT/A pharmacokinetics and pharmacodynamics: entering the body, distribution, receptor binding, translocation, mediator release suppression, zinc metabolism as well as factors affecting body sensitivity to BoNT at each of those stages. The specific biological effects of BoNT/A, which may underlie its analgesic, anticancer and anti-inflammatory effects, are described. Botulinum neurotoxin pharmacokinetics and pharmacodynamics features discussed herein represent significant clinical relevance since they determine botulinum treatment safety and effectiveness. And also they open ways to develop both BoNT-based therapies and anti-botulinic agents.

Keywords: botulinum, neurotoxin, ganglioside, synaptic vesicle protein, fibroblast growth factor receptor, SNARE proteins, mediator release block, zinc metabolism, thioredoxin reductase - thioredoxin system, botulism, neuromuscular blockade

1. Introduction

Botulinum neurotoxins (BoNTs) are the most potent protein toxins among bacterial, animal, plant and chemical toxic compounds and are the cause of botulism [1]. However, BoNT-based therapeutics are widely used for treatment of various diseases and esthetic disorders. Various features of BoNT pharmacology holding a great promise for development of both BoNT-based therapeutics and anti-botulinic agents are currently studied. Unique molecular mechanisms underlying various stages of botulinum neurotoxin type A pharmacological activity as well as potential factors affecting body sensitivity to BoNT are described herein.

2. Neurotoxin complex and BoNT molecule structure

Botulinum neurotoxin is a protein dimer with molecular weight of 150 kDa and chemical formula C6760H10447N1743O2010S32 consisting of two chains: light and heavy [1]. The light chain represents approximately one third of toxin molecular weight and is bound to the heavy one with a disulfide link [2].

The light chain (L-chain) is a protease blocking synaptic release. It forms the BoNT molecule catalytic domain. The heavy chain (H-chain) consists of two domains: binding domain bounds to target cell surface receptors, translocation domain is involved in light chain translocation creating cell membrane channel. The BoNT molecule is a dipole with an electric charge attenuating from the binding domain to the catalytic one [3]. It is of importance when the molecule is directed relative to cell membrane that facilitates receptor binding.

In natural settings BoNTs are synthesized by bacteria as a complex with several proteins: one non-toxic non-hemagglutinin (NTNHA) and several hemagglutinins [1].

NTNHA has a molecular weight of 130 kDa and its amino acid sequences are highly homologous to BoNT but without protease motif as the only difference. "Hand in glove"-type interaction with the BoNT molecule protects the one from aggressive effects of environmental factors including GIT proteolytic enzymes [4].

There are three classes of hemagglutinins with molecular weight of 33–35, 15–18 and 70 kDa [5]. They do not contact with the BoNT molecule directly but with NTNHA working as an adhesin molecule when such toxin complex is absorbed.

Non-toxic hemagglutinin and hemagglutinin proteins can form various multimeric complexes with BoNT called botulinum neurotoxin complexes. Each of them contains only one BoNT molecule released from the complex if medium pH changes [2].

3. BoNT absorption and distribution

BoNTs can enter human body via both injured and intact tissues. Therapeutically botulinum neurotoxin type-A (BoNT/A) based agents are mainly injected as close as possible to their target cells. However, BoNT/A forms to be applied without the need to damage skin are already under development though they are not yet through Phase III clinical studies [6, 7].

In natural settings BoNTs show systemic action causing botulism and enter the body mainly through intact membranes.

Depending on toxin mode of entry botulism forms can be classified as follows: food botulism (ingestion of BoNT-contaminated food), infant (ingestion of food with bacteria spores), inhalation (breathing-in BoNT-containing aerosols), wound (in majority of cases it is related to injectable drug use), iatrogenic [8].

In natural settings botulinum neurotoxin should cross epithelial barriers and reach general circulation to hit its target cells. Such process is called absorption.

BoNT might utilize two modes of penetrating intestinal or pulmonary epithelium: intracellular route and intercellular junction-related one.

In case of transcytosis (penetration through epithelial cell) BoNT binds to ganglioside receptors at epithelial cell surface and undergoes endocytosis (being captured in a vesicle). Transport vesicles transfer toxin through the whole cell and release it into general circulation. Neither toxin structure is altered, nor it is released in cell cytosol during transcytosis, which differentiates BoNT binding with epithelial cell from binding with neuronal ones [9, 10].

Paracellular route (through intercellular junctions) may or may not involve complexing proteins. Complexin hemagglutinins can bind to E-cadherin in epithelial intercellular junctions and disrupts the latter allowing BoNT in general circulation [4]. However, BoNT molecules are able to break epithelial barriers without complexing proteins.

Studies by Maksymowych et al. [11] and Al-Saleem et al. [12] showed that introduction of equimolar amounts of free BoNT/A and BoNT/A complexes resulted in equivalent BoNT titers in general circulation with similar toxicity and effectiveness. However, hemagglutinins are assumed to boost BoNT transportation through epithelium.

When transported through intestinal wall BoNT may bind to cholinergic and serotoninergic neurons of enteral (intestinal) nervous system located in intestinal submucosa blocking gut motor and secretory activity. It explains impaired bowel movement (constipation) as one of early signs for alimentary and infant botulism [13].

BoNT penetrating epithelial barrier reaches general circulation and is distributed in all extracellular fluid compartments in the body but the ones of central nervous system.

Eisele et al. [14] had a series of experiments demonstrating that with pH values close to neutral (arterial blood pH of 7.37–7.43 [15, 16]) botulinum neurotoxin complex dissociates on active BoNT and complexing proteins with half-life below 1 minute. Once such toxin complex dissociates complexing proteins are not any more of any significance for the occurrence of the clinical effect of BoNT.

Al-Saleem et al. [17] works proved that toxin reaches general circulation without any evident structural or biological activity changes. General circulation performs as toxin storage compartment until BoNT reaches its target cells. While in general circulation BoNT undergoes slight biotransformation, it is not accumulated in blood cells and mostly remains in its free active form. Such concept of "general circulation—botulinum neurotoxin storage compartment" has been confirmed by many researchers. Fagan et al. [18] described active BoNT/A presence in human blood serum 11 days after contaminated food ingestion; Sheth et al. [19], 25 days after disease onset; Delbrassinne et al. [20], 29 days after contaminated food was taken.

From intravascular fluid compartment botulinum neurotoxin enters extravascular one and then intercellular fluid. Being locally injected with therapeutic purposes botulinum neurotoxin is directly introduced in extravascular compartment (or intravascular one if it is in a blood vessel) next to target cells bypassing absorption stage. From intercellular compartment botulinum neurotoxin should reach its target—peripheral cholinergic nerve endings and bind to receptors there.

To better understand the mechanism of botulinum neurotoxin binding with receptors knowledge of normal neurotransmission in synapses is required.

4. Normal synapse neurotransmission

Neuromediators are synthesized in neuron cytosol and then stored in pre-synaptic nerve endings within synaptic vesicles. Synaptic vesicle membrane contains proton pump (vesicular ATPase), which, when activated, increases intravesicular proton concentration [8]. Electrochemical proton gradient ensures mediator influx from cytosol and its accumulation in such vesicles. The uptake of mediators within the synaptic vesicles is also regulated by receptors on the neuronal membrane, not only by the proton gradient. Mediator-containing vesicles are located in neuron cytoplasm and are bound to specific presynaptic membrane regions (active zones [21]) during so-called docking [22]. Vesicles are docked with cell membrane in active zones only and docking is controlled by a great deal of transport proteins [23].

When a nerve impulse arrives axonal presynaptic membrane is depolarized, calcium channels open and Ca^{2+} ions flow into axon [24]. In response to Ca^{2+} influx the mediator-containing vesicle fuses with presynaptic membrane in active zone. This stage is called priming. It is regulated with two integral membrane synaptic vesicle proteins (synaptobrevin and synaptotagmin) as well as two presynaptic membrane proteins (SNAP25 and syntaxin) and cytosol proteins including complexin [25–28].

Rapid vesicle conformation changes by regulatory proteins result in full synaptic vesicle fusion with presynaptic membrane and pore formation wherethrough neuromediator is released in a synaptic cleft [29].

Neuromediator diffuses from its nerve terminal and binds to post-synaptic receptors that trigger post-synaptic cell signaling. In neuromuscular junctions acetyl choline binds with its receptor on myocyte plasmalemma resulting in muscle cell membrane depolarization. Membrane depolarization kicks off Ca^{2+} influx in myocyte and muscle contraction.

While neuromediator is released synaptic vesicle lumen opens temporarily into a synaptic cleft but later it internalizes in nerve terminal during endocytosis. After endocytosis the vesicle is again filled in with neuromediator and next neurotransmission cycle starts [30].

5. BoNT/A binding with target cells

Active BoNT/A molecules bind with target cells via their receptors on cell surface [31]. To bind with neuronal membrane BoNT/A molecule should interact with a set of high and low affinity receptors [32]. Currently three receptors (polysyaloganglioside GT1b, fibroblast growth factor receptor 3, transmembrane vesicular receptor SV2) and several co-receptors have been described with such combination.

Active neurotoxin molecule endocytosis and its further changes are possible only once it binds with entire receptor combination at axonal surface [5]. Binding to one of the receptors without interaction with others does not induce toxin internalization. Such multistage process for BoNT/A binding with receptors makes up for low BoNT/A concentration in circulating fluids, high rate of extracellular flow around cells and small axonal surface area.

5.1 First receptor: Polysyaloganglioside

First BoNT/A receptor at neuronal surface is polysyaloganglioside GT1b (PSG).

Gangliosides are glycosylated lipids being a part of cell membranes. Though gangliosides are present in all tissues of vertebrates they are more prevalent in neuronal membranes [33] where they are involved in optimal myelin production, axon-myelin interactions, peripheral and central axon stability [32].

PSG density on presynaptic membrane is high. PSGs are grouped as microdomains next to presynaptic membrane active zones [34]. PSG receptor presence in these zones is important for processes of botulinum neurotoxin binding with other receptors.

Oligosaccharide (BoNT-binding part) PSG projects quite far outside membrane surface in a synaptic gap and is negatively charged [8]. BoNT/A molecule is a dipole with positively charged binding domain [3]. Such electric charge difference of BoNT/A binding domain and PSG receptors (and other anion lipids at axonal membrane) makes possible to redirect BoNT/A molecule on its way to cell membrane enhancing receptor binding chances.

Currently polysyaloganglioside are considered as initial binding regions drawing toxin from relatively vast 3D extracellular fluid space into 2D membrane surface one [5]. It is required, in turn, for toxin binding to following receptors. On one hand, binding to PSG is irreversible since BoNT/A is extracted from ground substance and is fixed on axonal membrane. On the other, at that stage toxin can still be affected and neutralizing antibodies can still reach it.

However, polysyalogangliosides are membrane receptors for both botulinum neurotoxin and human neuropathy-associated antiganglioside autoantibodies. Anti-PSG autoantibody production in neuropathy patients may induce diminished botulinum neurotoxin sensitivity and resistance development [35].

5.2 Second receptor: Fibroblast growth factor receptor 3

HC subdomain structure of botulinum neurotoxin type A is homologous to basic fibroblast growth factor (FGF) [36]. That similarity enables BoNT/A high-affinity binding with protein fibroblast growth factor receptor 3 (FGFR3b) on neuronal surface [37].

However, FGFR3b receptors are affine not only to BoNT/A but also to multiple fibroblast growth factors. Moreover, this receptor affinity to growth factors exceeds the one to botulinum neurotoxin. Native FGFR3 ligands—growth factors FGF1, FGF2 and FGF9—compete for binding with FGFR3 and occupying receptors are able to jam BoNT/A absorption by cells [8].

Besides, FGFR3b receptor activity is regulated by several low-affinity cofactors including heparansulfate, neuropilin-1, anosmin, etc. [38]. Non-specificity and competitive binding of FGFR3 receptors with BoNT/A and fibroblast growth factors, cofactor impact on receptor activity may explain fragility of the said receptor mechanism and, therefore, variable sensitivity to botulinum neurotoxin. Moreover, some FGFR3 mutation-related conditions (skeletal dysplasias, epidermal nevus, seborrheic keratosis, hyperinsulinemia) might demonstrate defective FGFR3 expression [39–43]. FGFR3 mutation influence on botulinum neurotoxin sensitivity is yet to be studied.

5.3 Third receptor: Transmembrane vesicular receptor SV2

SV2 is a protein receptor located on vesicular membrane [44] of all peripheral and central neurons as well as on secretory granule membrane of endocrine cells [45]. SV2 is expressed on vesicular membranes in cells accumulating not only acetyl choline but also GABA, dopamine, glutamate, substance P and several other mediators [46].

Unlike polysyaloganglioside receptors expressed into a synaptic gap the SV2-receptor BoNT/A-binding site is projected into synaptic vesicle lumen and is not approachable for neurotoxin while such vesicle is in axonal cytosol [47]. SV2 becomes reachable for BoNT/A at the time of vesicle fusion with presynaptic membrane and acetyl choline exocytosis [48].

Thus, BoNT/A binding with entire receptor combination happens in active zones only after synaptic vesicle fusion with presynaptic membrane and opening of vesicular lumen into synaptic cleft facilitating further BoNT/A endocytosis. After binding with receptor combination and endocytosis botulinum neurotoxin cannot be reached by neutralizing antibodies anymore.

6. Endocytosis

BoNT/A molecule binding with receptors results in receptor-mediated endocytosis of both receptors and toxin [49].

Immediately after endocytosis vesicular lumen has neutral pH. Vesicular ATPase proton pump controls mediator re-uptake [50] and injects protons into synaptic vesicle, therefore, gradually decreasing vesicular lumen pH [51].

7. Light chain translocation

Vesicular medium acidification results in irreversible conformation changes of both heavy and light BoNT/A chains. With these changes the heavy chain being linked via receptors with vesicular membrane forms transmembrane H-channel there [52, 53]. Through the channel the conformation-altered light chain leaves the vesicle for cytosol [54] and then chain-binding disulfide link breaks up.

L-chain translocation occurs with pH between 4.5 and 6 [55]. pH decrease results in protonation of carboxylated amino acid residues present in BoNT/A heavy and light chains. Carboxylated residues are located at one side of toxin molecule and their protonation results in significant changes of molecular shape [55]. BoNT/A molecule with its positively charged surface interacts with anion vesicular membrane surface forming protein and lipid complex [56]. L-chain is assumed to turn into "molten protein globule" gaining hydrophobic features [8]. On one hand, L-chain hydrophobicity ensures its translocation via the H-chain-formed membrane channel. On the other, with lower pH molecular surface where the disulfide bond is located becomes more hydrophobic. It ensures disulfide bond integrity until complete L-chain translocation.

To cross vesicular membrane L-chain should have disulfide bond with H-chain throughout entire translocation sequence [55]. Premature disulfide bond breakage at any stage until it exits into cytosol interrupts L-chain translocation [57].

At the end of translocation process the disulfide bond is destroyed by thioredoxin reductase-thioredoxin system releasing light chain to express its catalytic activity in cytosol [58].

Thioredoxin reductase (TrxR)—thioredoxin (Trx) system is a main cellular redox system. TrxR and Trx are cytosol side proteins of vesicular membrane and their inhibition may block BoNT/A action on stages when neurotoxin cannot be reached by neutralizing antibodies [59]. In vitro experiments of Zanetti et al. [60] showed that inhibitors for TrxR-Trx enzymatic couple hampers L-chain protease activity for all known botulinum neurotoxin serotypes in cultured neurons. While in vivo they prevent toxin-induced paralysis in mice irrespective of botulinum neurotoxin serotype.

In terms of life cycle model disulfide bond reduction is the end of intracellular existence of intact active BoNT/A molecule (holotoxin). Even if a light or heavy chain was exported out of cell none of them on its own should be able to disrupt cell functioning. Only holotoxin can undergo through multiple stages ended up with conduction block [61]. On the other side, conformation changes related to pH-induced L-chain translocation in cytosol create "a trap" making impossible both retrotranslocation into endosome and active toxin molecule return in extracellular environment [5].

8. L-chain cleaves transport proteins

Modified L-chain enters neuron cytosol through H-channel where it behaves as a metalloprotease. It catalytically cleaves nine amino acids from C-terminal of soluble

N-ethylmaleimide-sensitive factor-attachment protein receptor (SNARE) for SNAP25 protein (SNAP25206) forming SNAP25197 [62, 63]. Intact SNAP25 is required for mediator-containing vesicle attachment with further neurotransmitter release and it is also involved directly in Ca-channels activity regulation in presynaptic membrane [64]. SNAP25 cleavage impairs mediator exocytosis causing nerve impulse conduction block and muscle paralysis [65].

Synaptic activity is highly sensitive to cleavage of minimal SNAP25 amounts. It was hypothesized that SNAP25 in neuron cytosol exists as various pools and that only small amounts of SNAP25 are actively involved in exocytosis and reachable for L-protease effects [66]. It was confirmed experimentally showing that cleavage of 10–15% of total intracellular SNAP25 pool is sufficient for complete neuromediator release block [67–69]. L-protease cleavage of as little as 2–3% of SNAP25 pool results in block of miniature post-synaptic cell potentials (weak depolarization of post-synaptic membranes at neuromuscular rest) [70].

Along with that SNAP25 proteolysis product, SNAP25197 protein, on its own inhibits exocytosis [71]. Meunier et al [72]. described that SNAP25197 is able to persist for a long while in cytosol as a component of the non-productive SNARE complex prolongating BoNT/A effects. While removal of several amino acids from SNAP25197 results in rapid exocytosis restoration.

9. Zinc metabolism and translocation

Zinc is necessary for light chain catalytic activity. One botulinum holotoxin molecule contains 1 zinc atom retained by L-chain zinc-binding amino acid sequence and such binding is reversible [73].

Vesicle acidification causes protonation of zinc-binding sections in the BoNT molecule. Translocation causes light chain denaturation obliterating chelate site integrity. As a result bound zinc dissociates and adds up to cytosol zinc pool.

Simpson et al [74]. in their in vitro studies demonstrated that zinc removal from active botulinum neurotoxin molecule caused L-chain catalytic activity loss in cell-free samples. Though activity in intact neuromuscular junctions retained since internalized toxin bound cytosol zinc. Thus, zinc retained by holotoxin (intact active molecule) is not the same zinc that is bound with catalytically active light chain. Light chain binds cytosol zinc.

10. Mediator release block

Main BoNT/A target is peripheral neurons where botulinum neurotoxin inhibits acetylcholine release [75].

Many of cell-based studies showed that BoNT/A not only blocks acetylcholine release but also prevents release of multiple other neuromediators if they are accumulated and stored in vesicles [32]. These neuromediators are as follows: epinephrine, norepinephrine, dopamine [76, 77], glutamate [78], glycine [79], serotonin [80], substance P [81], etc. Therefore, botulinum neurotoxin is to be considered not as specific acetylcholine release inhibitor but rather as an exocytosis blocker for various mediators that offers tremendous promise for treatment and prevention of various disorders.

11. Specific biological effects

In addition, BoNT/A can affect cells not only as a blocker of exocytosis mediators, but also by binding to various receptors on the cell membrane, cause specific biological effects. Including influencing the expression of genes by the cell. Grando and Zachary [82] described that many cells are capable of expressing one or more BoNT/A receptors and binding BoNT/A: epidermal keratinocytes, mesenchymal stem cells from subcutaneous adipose tissue, nasal mucosa cells, urothelium, intestinal epithelial cells, prostate epithelial cells, alveolar epithelial cells, neutrophils, macrophages, etc. In addition to SNAP25, BoNT/A can also cleave SNAP-23, which is expressed in various human tissues.

Kim et al. [83] experimentally proved that BoNT/A is able to bind to TLR2 receptors on macrophages, changing the expression of genes responsible for signal transduction, protein metabolism and modification, nucleic acid metabolism, apoptosis, proliferation, cell differentiation. Which may explain the anti-inflammatory effect of BoNT/A.

12. Cytotoxicity

Cytotoxicity for BoNT/A has not been established either in cell-based studies [84] or in electrophysiological studies in healthy humans [85]. In addition, the experience of therapeutic use of BoNT/A for various indications indicates the absence of any signs of neuronal damage even with long-term regular use [66, 86, 87].

13. Conclusion

Further studies of unique pharmacological mechanisms for botulinum neurotoxin are quite promising for the search of the ways to influence its effects: to extend/shorten its action duration, to increase/decrease BoNT sensitivity in specific patient populations. Also it will help to develop protocols for optimal combinations of botulinum neurotoxin with esthetic medicine procedures of all kinds. Better insights on multiple aspects of not only BoNT neuronal selectivity but also BoNT/A interaction with non-neuronal cells will show ways to find new therapeutic applications of botulinum neurotoxin-based agents in various areas of medicine.

One of the promising areas of botulinum therapy and bioengineering of botulinum toxin is the treatment of pain syndromes. BoNT/A is effective in the treatment of various neuropathic pain syndromes, including chronic migraine, postherpetic neuralgia, trigeminal neuralgia and peripheral neuralgia [32].

Hybrid preparations of botulinum toxin are being developed to suppress secretion in various cell populations, including the secretion of growth hormone [88]. Various effects of BoNT/A on the enhancement and suppression of gene expression in neuronal and non-neuronal cells are described [89]. This implies a fundamentally different response of neuronal and epithelial cells to the action of botulinum toxin and is of great importance in the development of anti-cancer treatments based on BoNT/A.

Thus, BoNT/A should not be considered as a specific blocker of acetylcholine release by motor neurons, but mainly as a blocker of exocytosis of various mediators by various cells, neuronal and non-neuronal. Moreover, the biological effect of BoNT/A can be realized not only through the blockade of exocytosis. And it can be absolutely different in different types of cells, which has great prospects in the treatment and prevention of many diseases.

Author details

Anna V. Reznik
ARclinic Medical Center, Saint-Petersburg, Russia

*Address all correspondence to: ksho@yandex.ru

References

[1] Poulain B, Popoff M. Why are botulinum neurotoxin-producing bacteria so diverse and botulinum neurotoxins so toxic? Toxins (Basel). 2019;**11**(1):34. DOI: 10.3390/toxins11010034

[2] Berry M. Botulinum neurotoxin: Basic facts, physiology and pharmacology. Atlas of Surgical Therapy for Migraine and Tension-Type Headache. 2019:45-46. DOI: 10.1007/978-3-030-29505-9_7

[3] Fogolari F, Tosatto S, Muraro L, Montecucco C. Electric dipole reorientation in the interaction of botulinum neurotoxins with neuronal membranes. FEBS Letters. 2009;**583**(14):2321-2325. DOI: 10.1016/j.febslet.2009.06.046

[4] Gu S, Rumpel S, Zhou J, et al. Botulinum neurotoxin is shielded by NTNHA in an interlocked complex. Science. 2012;**335**(6071):977-981. DOI: 10.1126/science.1214270

[5] Simpson L. The life history of a botulinum toxin molecule. Toxicon. 2013;**68**:40-59. DOI: 10.1016/j.toxicon.2013.02.014

[6] Zhu Z, Stone H, Thach T, Garcia L, Ruegg C. A novel botulinum neurotoxin topical gel: Treatment of allergic rhinitis in rats and comparative safety profile. American Journal of Rhinology & Allergy. 2012;**26**(6):450-454. DOI: 10.2500/ajra.2012.26.3785

[7] Collins A, Nasir A. Topical Botulinum Toxin. The Journal of Clinical and Aesthetic Dermatology (JCAD). 2010. Available from: https://jcadonline.com/topical-botulinum-toxin/ [Accessed: 9 August 2020]

[8] Rossetto O, Pirazzini M, Montecucco C. Botulinum neurotoxins: genetic, structural and mechanistic insights. Nature Reviews Microbiology. 2014;**12**(8):535-549. DOI: 10.1038/nrmicro3295

[9] Couesnon A, Shimizu T, Popoff M. Differential entry of botulinum neurotoxin A into neuronal and intestinal cells. Cellular Microbiology. 2009;**11**(2):289-308. DOI: 10.1111/j.1462-5822.2008.01253.x

[10] Elias M, Al-Saleem F, Ancharski D, et al. Evidence that botulinum toxin receptors on epithelial cells and neuronal cells are not identical: Implications for development of a non-neurotropic vaccine. Journal of Pharmacology and Experimental Therapeutics. 2010;**336**(3):605-612. DOI: 10.1124/jpet.110.175018

[11] Maksymowych A, Reinhard M, Malizio C, Goodnough M, Johnson E, Simpson L. Pure botulinum neurotoxin is absorbed from the stomach and small intestine and produces peripheral neuromuscular blockade. Infection and Immunity. 1999;**67**(9):4708-4712. DOI: 10.1128/iai.67.9.4708-4712.1999

[12] Al-Saleem F, Ancharski D, Joshi S, et al. Analysis of the mechanisms that underlie absorption of botulinum toxin by the inhalation route. Infection and Immunity. 2012;**80**(12):4133-4142. DOI: 10.1128/iai.00669-12

[13] Rosow L, Strober J. Infant botulism: Review and clinical update. Pediatric Neurology. 2015;**52**(5):487-492. DOI: 10.1016/j.pediatrneurol.2015.01.006

[14] Eisele K, Fink K, Vey M, Taylor H. Studies on the dissociation of botulinum neurotoxin type A complexes. Toxicon. 2011;**57**(4):555-565. DOI: 10.1016/j.toxicon.2010.12.019

[15] Boyle J, Weitzman J, Berne C. Indications for measurement of arterial blood pH. The American Journal of Surgery. 1960;**100**(2):346-353. DOI: 10.1016/0002-9610(60)90308-1

[16] Kaplan L, Kellum J. Fluids, pH, ions and electrolytes. Current Opinion in Critical Care. 2010;**16**(4):323-331. DOI: 10.1097/mcc.0b013e32833c0957

[17] Al-Saleem F, Ancharski D, Ravichandran E, et al. The role of systemic handling in the pathophysiologic actions of botulinum toxin. Journal of Pharmacology and Experimental Therapeutics. 2008;**326**(3):856-863. DOI: 10.1124/jpet.108.136242

[18] Fagan R, McLaughlin J, Middaugh J. Persistence of botulinum toxin in patients'serum: Alaska, 1959-2007. The Journal of Infectious Diseases. 2009;**199**(7):1029-1031. DOI: 10.1086/597310

[19] Sheth A, Wiersma P, Atrubin D, et al. International outbreak of severe botulism with prolonged toxemia caused by commercial carrot juice. Clinical Infectious Diseases. 2008;**47**(10):1245-1251. DOI: 10.1086/592574

[20] Delbrassinne L, Laisnez V, De Weweire M, Vanderpas J, Dierick K, Denayer S. Very long persistence of botulinum toxin B in a patient's serum. Open Infect Dis J. 2018;**10**(1):187-191. DOI: 10.2174/1874279301810010187

[21] Zhai R, Bellen H. The architecture of the active zone in the presynaptic nerve terminal. Physiology. 2004;**19**(5):262-270. DOI: 10.1152/physiol.00014.2004

[22] Heuser J, Reese T. Evidence for recycling of synaptic vesicle membrane during transmitter release at the frog neuromuscular junction. The Journal of Cell Biology. 1973;**57**(2):315-344. DOI: 10.1083/jcb.57.2.315

[23] Ahmari S, Buchanan J, Smith S. Assembly of presynaptic active zones from cytoplasmic transport packets. Nature Neuroscience. 2000;**3**(5):445-451. DOI: 10.1038/74814

[24] Stanley E. The calcium channel and the organization of the presynaptic transmitter release face. Trends in Neurosciences. 1997;**20**(9):404-409. DOI: 10.1016/s0166-2236(97)01091-6

[25] Catterall W. Interactions of presynaptic ca2+ channels and snare proteins in neurotransmitter release. Annals of the New York Academy of Sciences. 1999;868(1 MOLECULAR AND):144-159. DOI: 10.1111/j.1749-6632.1999.tb11284.x

[26] Hilfiker S, Pieribone V, Czernik A, Kao H, Augustine G, Greengard P. Synapsins as regulators of neurotransmitter release. Philosophical Transactions of the Royal Society of London Series B: Biological Sciences. 1999;**354**(1381):269-279. DOI: 10.1098/rstb.1999.0378

[27] Hallam S, Goncharov A, McEwen J, Baran R, Jin Y. SYD-1, a presynaptic protein with PDZ, C2 and rhoGAP-like domains, specifies axon identity in C. elegans. Nature Neuroscience. 2002;**5**(11):1137-1146. DOI: 10.1038/nn959

[28] Jarvis S, Barr W, Feng Z, Hamid J, Zamponi G. Molecular determinants of syntaxin 1 modulation of N-type calcium channels. Journal of Biological Chemistry. 2002;**277**(46):44399-44407. DOI: 10.1074/jbc.m206902200

[29] Taverna E, Saba E, Rowe J, Francolini M, Clementi F, Rosa P. Role of lipid microdomains in P/Q-type calcium

channel (Cav2.1) clustering and function in presynaptic membranes. Journal of Biological Chemistry. 2003;**279**(7):5127-5134. DOI: 10.1074/jbc.m308798200

[30] Burns M, Augustine G. Synaptic structure and function: Dynamic organization yields architectural precision. Cell. 1995;**83**(2):187-194. DOI: 10.1016/0092-8674(95)90160-4

[31] Simpson L. Molecular pharmacology of botulinum toxin and tetanus toxin. Annual Review of Pharmacology and Toxicology. 1986;**26**(1):427-453. DOI: 10.1146/annurev.pa.26.040186.002235

[32] Poulain B, Lemichez E, Popoff M. Neuronal selectivity of botulinum neurotoxins. Toxicon. 2020;**178**:20-32. DOI: 10.1016/j.toxicon.2020.02.006

[33] Schnaar R. Gangliosides of the vertebrate nervous system. Journal of Molecular Biology. 2016;**428**(16):3325-3336. DOI: 10.1016/j.jmb.2016.05.020

[34] Prinetti A, Loberto N, Chigorno V, Sonnino S. Glycosphingolipid behaviour in complex membranes. Biochimica et Biophysica Acta (BBA) - Biomembranes. 2009;**1788**(1):184-193. DOI: 10.1016/j.bbamem.2008.09.001

[35] Bullens R, O'Hanlon G, Wagner E, et al. Complex gangliosides at the neuromuscular junction are membrane receptors for autoantibodies and botulinum neurotoxin but redundant for normal synaptic function. The Journal of Neuroscience. 2002;**22**(16):6876-6884. DOI: 10.1523/jneurosci.22-16-06876.2002

[36] Jacky B, Garay P, Dupuy J, et al. Identification of fibroblast growth factor receptor 3 (FGFR3) as a protein receptor for botulinum neurotoxin serotype a (BoNT/A). PLoS Pathogens. 2013;**9**(5):e1003369. DOI: 10.1371/journal.ppat.1003369

[37] James NG, Malik S, Sanstrum BJ, et al. Characterization of clostridium botulinum neurotoxin serotype A (BoNT/A) and fibroblast growth factor receptor interactions using novel receptor dimerization assay. Sci Rep. 2021;**11**:7832. DOI: 10.1038/s41598-021-87331-7

[38] Rummel A. Double receptor anchorage of botulinum Neurotoxins accounts for their exquisite neurospecificity. In: Rummel A., Binz T. (eds) Botulinum Neurotoxins. Current Topics in Microbiology and Immunology. Vol. 364. Berlin, Heidelberg: Springer; 2012:61-90. DOI: 10.1007/978-3-642-33570-9_4

[39] Hernández S, Toll A, Baselga E, et al. Fibroblast growth factor receptor 3 mutations in epidermal nevi and associated low grade bladder tumors. Journal of Investigative Dermatology. 2007;**127**(7):1664-1666. DOI: 10.1038/sj.jid.5700705

[40] Hafner C, Hartmann A, Vogt T. FGFR3 mutations in epidermal nevi and seborrheic keratoses: Lessons from urothelium and skin. Journal of Investigative Dermatology. 2007;**127**(7):1572-1573. DOI: 10.1038/sj.jid.5700772

[41] Chen J, Liu J, Zhou Y, et al. Molecular therapeutic strategies for FGFR3 gene-related skeletal dysplasia. Journal of Molecular Medicine. 2017;**95**(12):1303-1313. DOI: 10.1007/s00109-017-1602-9

[42] Mustafa M, Moghrabi N, Bin-Abbas B. Hypochondroplasia, acanthosis nigricans, and insulin resistance in a child with FGFR3 mutation: Is it just an association? Case Rep Endocrinol. 2014;**2014**:1-6. DOI: 10.1155/2014/840492

[43] Blomberg M, Jeppesen E, Skovby F, Benfeldt E. FGFR3 mutations and the

skin: Report of a patient with a FGFR3 gene mutation, acanthosis nigricans, hypochondroplasia and hyperinsulinemia and review of the literature. Dermatology. 2010;**220**(4):297-305. DOI: 10.1159/000297575

[44] Dong M. SV2 is the protein receptor for botulinum neurotoxin A. Science. 2006;**312**(5773):592-596. DOI: 10.1126/science.1123654

[45] Bartholome O, Van den Ackerveken P, Sánchez Gil J, et al. Puzzling out synaptic vesicle 2 family members functions. Frontiers in Molecular Neuroscience. 2017;**10**:148. DOI: 10.3389/fnmol.2017.00148

[46] Dunn A, Hoffman C, Stout K, Ozawa M, Dhamsania R, Miller G. Immunochemical analysis of the expression of SV2C in mouse, macaque and human brain. Brain Research. 2019;**1702**:85-95. DOI: 10.1016/j.brainres.2017.12.029

[47] Strotmeier J, Mahrhold S, Krez N, et al. Identification of the synaptic vesicle glycoprotein 2 receptor binding site in botulinum neurotoxin A. FEBS Letters. 2014;**588**(7):1087-1093. DOI: 10.1016/j.febslet.2014.02.034

[48] Yao G, Zhang S, Mahrhold S, et al. N-linked glycosylation of SV2 is required for binding and uptake of botulinum neurotoxin A. Nature Structural & Molecular Biology. 2016;**23**(7):656-662. DOI: 10.1038/nsmb.3245

[49] Colasante C, Rossetto O, Morbiato L, Pirazzini M, Molgó J, Montecucco C. Botulinum neurotoxin type A is internalized and translocated from small synaptic vesicles at the neuromuscular junction. Molecular Neurobiology. 2013;**48**(1):120-127. DOI: 10.1007/s12035-013-8423-9

[50] Takamori S, Holt M, Stenius K, et al. Molecular anatomy of a trafficking organelle. Cell. 2006;**127**(4):831-846. DOI: 10.1016/j.cell.2006.10.030

[51] Saheki Y, De Camilli P. Synaptic vesicle endocytosis. Cold Spring Harbor Perspectives in Biology. 2012;**4**(9):a005645-a005645. DOI: 10.1101/cshperspect.a005645

[52] Montecucco C, Schiavo G. Structure and function of tetanus and botulinum neurotoxins. Quarterly Reviews of Biophysics. 1995;**28**(4):423-472. DOI: 10.1017/s0033583500003292

[53] Montal M. Translocation of botulinum neurotoxin light chain protease by the heavy chain protein-conducting channel. Toxicon. 2009;**54**(5):565-569. DOI: 10.1016/j.toxicon.2008.11.018

[54] Brunger A, Breidenbach M, Jin R, Fischer A, Santos J, Montal M. Botulinum neurotoxin heavy chain belt as an intramolecular chaperone for the light chain. PLoS Pathogens. 2007;**3**(9):e113. DOI: 10.1371/journal.ppat.0030113

[55] Pirazzini M, Rossetto O, Bolognese P, Shone C, Montecucco C. Double anchorage to the membrane and intact inter-chain disulfide bond are required for the low pH induced entry of tetanus and botulinum neurotoxins into neurons. Cellular Microbiology. 2011;**13**(11):1731-1743. DOI: 10.1111/j.1462-5822.2011.01654.x

[56] Galloux M, Vitrac H, Montagner C, et al. Membrane interaction of botulinum neurotoxin A translocation (T) domain. Journal of Biological Chemistry. 2008;**283**(41):27668-27676. DOI: 10.1074/jbc.m802557200

[57] Fischer A, Montal M. Crucial role of the disulfide bridge between

botulinum neurotoxin light and heavy chains in protease translocation across membranes. Journal of Biological Chemistry. 2007;**282**(40):29604-29611. DOI: 10.1074/jbc.m703619200

[58] Pirazzini M, Azarnia Tehran D, Zanetti G, Rossetto O, Montecucco C. Hsp90 and thioredoxin-thioredoxin reductase enable the catalytic activity of clostridial neurotoxins inside nerve terminals. Toxicon. 2018;**147**:32-37. DOI: 10.1016/j.toxicon.2017.10.028

[59] Pirazzini M, Azarnia Tehran D, Zanetti G, et al. Thioredoxin and its reductase are present on synaptic vesicles, and their inhibition prevents the paralysis induced by botulinum neurotoxins. Cell Reports. 2014;**8**(6):1870-1878. DOI: 10.1016/j.celrep.2014.08.017

[60] Zanetti G, Azarnia Tehran D, Pirazzini M, et al. Inhibition of botulinum neurotoxins interchain disulfide bond reduction prevents the peripheral neuroparalysis of botulism. Biochemical Pharmacology. 2015;**98**(3):522-530. DOI: 10.1016/j.bcp.2015.09.023

[61] Fischer A, Montal M. Single molecule detection of intermediates during botulinum neurotoxin translocation across membranes. Proceedings of the National Academy of Sciences. 2007;**104**(25):10447-10452. DOI: 10.1073/pnas.0700046104

[62] Blasi J, Chapman E, Link E, et al. Botulinum neurotoxin A selectively cleaves the synaptic protein SNAP-25. Nature. 1993;**365**(6442):160-163. DOI: 10.1038/365160a0

[63] Fernandez-Salas E, Steward L, Ho H, et al. Plasma membrane localization signals in the light chain of botulinum neurotoxin. Proceedings of the National Academy of Sciences. 2004;**101**(9):3208-3213. DOI: 10.1073/pnas.0400229101

[64] Pozzi D, Corradini I, Matteoli M. The control of neuronal calcium homeostasis by SNAP-25 and its impact on neurotransmitter release. Neuroscience. 2019;**420**:72-78. DOI: 10.1016/j.neuroscience.2018.11.009

[65] Kasai H, Takahashi N, Tokumaru H. Distinct initial SNARE configurations underlying the diversity of exocytosis. Physiological Reviews. 2012;**92**(4):1915-1964. DOI: 10.1152/physrev.00007.2012

[66] Pirazzini M, Rossetto O, Eleopra R, Montecucco C. Botulinum neurotoxins: Biology, pharmacology, and toxicology. Pharmacological Reviews. 2017;**69**(2):200-235. DOI: 10.1124/pr.116.012658

[67] Raciborska D, Trimble W, Charlton M. Presynaptic protein interactionsin vivo: Evidence from botulinum A, C, D and E action at frog neuromuscular junction. European Journal of Neuroscience. 1998;**10**(8):2617-2628. DOI: 10.1046/j.1460-9568.1998.00270.x

[68] Keller J, Cai F, Neale E. Uptake of botulinum neurotoxin into cultured neurons. Biochemistry. 2004;**43**(2):526-532. DOI: 10.1021/bi0356698

[69] Montecucco C, Schiavo G, Pantano S. SNARE complexes and neuroexocytosis: How many, how close? Trends in Biochemical Sciences. 2005;**30**(7):367-372. DOI: 10.1016/j.tibs.2005.05.002

[70] Beske P, Scheeler S, Adler M, McNutt P. Accelerated intoxication of GABAergic synapses by botulinum neurotoxin A disinhibits stem cell-derived neuron networks prior to network silencing. Frontiers in Cellular Neuroscience. 2015;**9**:159. DOI: 10.3389/fncel.2015.00159

[71] Huang X. Truncated SNAP-25 (1-197), like botulinum neurotoxin A, can inhibit insulin secretion from HIT-T15 insulinoma cells. Molecular Endocrinology. 1998;**12**(7):1060-1070. DOI: 10.1210/me.12.7.1060

[72] Meunier F, Lisk G, Sesardic D, Dolly J. Dynamics of motor nerve terminal remodeling unveiled using SNARE-cleaving botulinum toxins: The extent and duration are dictated by the sites of SNAP-25 truncation. Molecular and Cellular Neuroscience. 2003;**22**(4):454-466. DOI: 10.1016/s1044-7431(02)00016-7

[73] Schiavo G, Rossetto O, Benfenati F, Poulain B, Montecucco C. Tetanus and botulinum neurotoxins are zinc proteases specific for components of the neuroexocytosis apparatus. Annals of the New York Academy of Sciences. 1994;710(1 Toxins and Ex):65-75. DOI: 10.1111/j.1749-6632.1994.tb26614.x

[74] Simpson L, Maksymowych A, Hao S. The role of zinc binding in the biological activity of botulinum toxin. Journal of Biological Chemistry. 2001;**276**(29):27034-27041. DOI: 10.1074/jbc.m102172200

[75] Pantano S, Montecucco C. The blockade of the neurotransmitter release apparatus by botulinum neurotoxins. Cellular and Molecular Life Sciences. 2013;**71**(5):793-811. DOI: 10.1007/s00018-013-1380-7

[76] Ashton A, Dolly J. Characterization of the inhibitory action of botulinum neurotoxin type A on the release of several transmitters from rat cerebrocortical synaptosomes. Journal of Neurochemistry. 1988;**50**(6):1808-1816. DOI: 10.1111/j.1471-4159.1988.tb02482.x

[77] Maisey E, Wadsworth J, Poulain B, et al. Involvement of the constituent chains of botulinum neurotoxins A and B in the blockade of neurotransmitter release. European Journal of Biochemistry. 1988;**177**(3):683-691. DOI: 10.1111/j.1432-1033.1988.tb14423.x

[78] Foran P, Mohammed N, Lisk G, et al. Evaluation of the therapeutic usefulness of botulinum neurotoxin B, C1, E, and F compared with the long lasting type A. Journal of Biological Chemistry. 2002;**278**(2):1363-1371. DOI: 10.1074/jbc.m209821200

[79] Neale E, Bowers L, Jia M, Bateman K, Williamson L. Botulinum neurotoxin a blocks synaptic vesicle exocytosis but not endocytosis at the nerve terminal. The Journal of Cell Biology. 1999;**147**(6):1249-1260. DOI: 10.1083/jcb.147.6.1249

[80] Najib A, Pelliccioni P, Gil C, Aguilera J. Clostridium neurotoxins influence serotonin uptake and release differently in rat brain synaptosomes. Journal of Neurochemistry. 2008;**72**(5):1991-1998. DOI: 10.1046/j.1471-4159.1999.0721991.x

[81] Lucioni A, Bales G, Lotan T, McGehee D, Cook S, Rapp D. Botulinum toxin type A inhibits sensory neuropeptide release in rat bladder models of acute injury and chronic inflammation. BJU International. 2008;**101**(3):366-370. DOI: 10.1111/j.1464-410x.2007.07312.x

[82] Grando SA, Zachary CB. The non-neuronal and nonmuscular effects of botulinum toxin: An opportunity for a deadly molecule to treat disease in the skin and beyond. British Journal of Dermatology. 2018;**178**(5):1011-1019. DOI: 10.1111/bjd.16080

[83] Kim Y, Kim J, Lee K et al. Botulinum neurotoxin type A induces TLR2-mediated inflammatory

responses in macrophages. PLoS One. 2015;**10**(4):e0120840. DOI: 10.1371/journal.pone.0120840

[84] Williamson LC, Neale EA. Syntaxin and 25-kDa synaptosomal-associated protein: differential effects of botulinum neurotoxins C1 and A on neuronal survival. Journal of Neuroscience Research. 1998;**52**(5):569-583. DOI: 10.1002/(SICI)1097-4547(19980601)52:5<569::AID-JNR9>3.0.CO;2-A

[85] Eleopra R, Tugnoli V, Quatrale R, et al. Botulinum neurotoxin serotypes A and C do not affect motor units survival in humans: An electrophysiological study by motor units counting. Clinical Neurophysiology. 2002;**113**(8):1258-1264. DOI: 10.1016/s1388-2457(02)00103-7

[86] Cohen JL, Scuderi N. Safety and patient satisfaction of AbobotulinumtoxinA for aesthetic use: A systematic review. Aesthetic Surgery Journal. 2017;**37**(suppl 1):S32-S44. DOI: 10.1093/asj/sjx010

[87] Sätilä H. Over 25 years of pediatric botulinum toxin treatments: What have we learned from injection techniques, doses, dilutions, and recovery of repeated injections? Toxins. 2020;**12**:440. DOI: 10.3390/toxins12070440

[88] Dressler D. Botulinum toxin drugs: Brief history and outlook. Journal of Neural Transmission. 2016;**123**:277-279. DOI: 10.1007/s00702-015-1478-1

[89] Thirunavukkarasusx N et al. Microarray analysis of differentially regulated genes in human neuronal and epithelial cell lines upon exposure to type A botulinum neurotoxin. Biochemical and Biophysical Research Communications. 2011;**405**(4):684-690. DOI: 10.1016/j.bbrc.2011.01.102

Chapter 4

Botulinum Neurotoxin Uses in Overactive Bladder

Mohamed H. Zahran, Ali Abdel Raheem, Ibrahim Alowidah and Diaa-Eldin Taha

Abstract

Botulinum neurotoxin has been approved for use in different urologic disorders, especially overactive bladder (OAB). OAB is highly prevalent, with a relevant impact on patients' quality of life and the available health resources. The prevalence of OAB is 11.8% with no significant difference between male and female. Botulinum neurotoxin is now approved as a treatment of choice for refractory overactive bladder (ROAB) after the failure of behavioral and pharmacological therapy. It is associated with improvement of urgency and urge urinary incontinence in 60–70% of patients. Yet the effect is not long-standing and fades out in a mean of 6-months duration and repeated injection is warranted. Moreover, its associated side effects are not uncommon, especially urinary tract infection and urine retention. New modifications have been studied to make it less invasive, effective, and tolerable by the patients through injection-free mode. The subject to be explained in the book chapter is the role of botulinum neurotoxin in ROAB, including the mechanism of action, different types of botulinum toxin used, the accepted dose, associated side effects, and comparison of the outcome to other available treatment modalities. In addition, a close look at the new accepted approaches for intravesical administration of botulinum toxin in the bladder will be done.

Keywords: botulinum neurotoxin, overactive bladder, refractory overactive bladder, outcomes

1. Introduction

The International Continence Society (ICS)/International Urogynecological Association (IUGA) defines overactive bladder (OAB) as urgency with or without urge incontinence (UUI), associated with urinary frequency and nocturia in the absence of pathological (e.g. UTI, stones, bladder tumor) and metabolic factors (e.g. diabetes) [1]. The diagnosis is made by exclusion. Whether there is nerve damage or not, OAB might be classified as neurogenic or idiopathic. The prevalence of OAB is 11.8% with no significant difference between male and female and the incidence increases with age [2]. The prevalence of OAB in adults aged ≥18 years was 16% in men and 16.9% in women in the USA, and in adults aged ≥40 years, it was 15.6% for men and 17.4% for women in Europe. In Asia, the prevalence of OAB was lower, but it was still 6.0% with no differences between male and female [3].

As a result, proper understanding and management of OAB is mandatory to improve patients' quality of life and decrease its socioeconomic burden. Behavioral therapy, bladder retraining, and pelvic floor exercise represent the first line of management of OAB. Pharmacotherapy is the second line of treatment. Anticholinergics and β3 agonists have been shown to be clinically effective in people with OAB [4]. Yet, it has many side effects such as dry mouth, dry eyes, constipation, blurred vision, dyspepsia, urine retention, and reduced cognitive function, which limit their use especially in elderly. More than 70% of patients discontinue medication within 6 months to 3 years due to side effects [5]. Refractory OAB (ROAB) develops when both behavioral therapy and oral medications become no longer effective [5]. The AUA/SUFU guidelines describe ROAB as a failure of behavioral therapy after 8–12 weeks and failure of at least one anticholinergic agent used for 4–8 weeks [6]. ROAB has an unknown prevalence rate, but it is not uncommon among the OAB population [7]. Third-line treatment should be considered when patients fulfill the ROAB diagnostic criteria. Sacral neuromodulation (SNM) and Botulinum neurotoxin-A (BoNT-A) have been recently demonstrated to be successful, with success rates reaching up to 60% and 70%, respectively [6].

BoNT-A is frequently used for cosmetic purposes and is used to treat strabismus, blepharospasm, muscle dystonia, hyperhidrosis, and migraine [8]. Carpenter was the first to prove that botulinum neurotoxin inhibits bladder contractility in rats in 1967. In 2011, BoNT-A was licensed for the treatment of urine incontinence (UI) caused by neurogenic overactive detrusors. The US Food and Drug Administration (FDA) approved it for the treatment of OAB in 2013 [9].

2. Pathophysiology and management of OAB

OAB is a chronic condition of urgency, frequency, nocturia with or without UUI. It is sub-classified into dry and wet type based on the absence or presence of the UUI. The exact cause of OAB is not well-understood. The pathophysiology varies between neurogenic, myogenic, or idiopathic factors [10]. The imbalance between the excitatory and inhibition neural pathway to the bladder is one of the underlying mechanisms. Also, the increased sensitivity of bladder muscle receptors and muscarinic receptor upregulation plays a role in OAB pathophysiology [11]. Another potential cause is the myogenic dysfunction secondary to structural or functional alteration of the detrusor smooth muscle. Idiopathic detrusor instability of undetermined underlying cause is another mechanism [10].

Proper history taking and at least a 3-day bladder diary are indicated for initial evaluation of patients in order to quantify OAB symptoms. Urodynamic evaluation is essential for establishing the diagnosis. The hallmark urodynamic feature of OAB is detrusor overactivity (DO). Yet, it may not be demonstrated in some patients due to the inability to reproduce symptoms during the urodynamic [12]. It is worth mentioning that urodynamic diagnosis has no proven predictive value for the treatment response [13]. And EAU guidelines recommended against routine urodynamic evaluation before starting the first line of treatment for uncomplicated OAB [14].

The first line of management is conservative treatment in the form of lifestyle modification (fluid restriction, decrease caffeine intake, weight reduction, and stop smoking), behavioral therapy, bladder retraining, timed voiding, and pelvic floor muscle exercise [14]. Bladder retraining and timed voiding work by setting a target time for using the toilet before it patient should not void. Once this is achieved the time can be lengthened. Thus the central control can be re-learned as in infancy. Pelvic floor muscle training; described by Kegel, aims at strengthening and rehabilitating the pelvic floor muscle, increasing its tone, and increasing urethral resistance [15].

The second recommended line of treatment is pharmacotherapy. It can be initiated together with conservative treatment or postponed till the failure of conservative treatment based on the OAB symptoms severity. The EAU and AUA guidelines recommended anticholinergics and β3 agonists as a pharmacological treatment of OAB [6, 14]. Anticholinergics are competitive muscarinic receptors antagonists. They prevent cholinergic muscarinic receptors activation in the urinary bladder and consequently reduce spontaneous detrusor muscle activity during the filling phase and decrease detrusor pressure. Many anticholinergics have been used in clinical trials and none proved to be superior to the others in OAB management. Dose escalation may be appropriate in certain patients and increases the response [16]. The efficacy of anticholinergics varies between 50% and 75%. They help to reduce urgency and UUI episodes along with reducing frequency of micturition. However, adherence to anticholinergic treatment is low and decreases over time because of lack of efficacy, adverse events, and/or cost and a significant number of patients will stop anticholinergic agents within the first 3 months [17, 18].

When anticholinergics are ineffective, non-tolerated, or contraindicated, β3 agonist (mirabegron) can be used. It activates the β-3 adrenergic receptor in the detrusor muscle in the bladder, which leads to muscle relaxation and an increase in bladder capacity helping the bladder to fill and store urine. Yet, it is not free of side effects. Tachycardia, hypertension, dyspepsia, palpitations, atrial fibrillation, joint swelling, rash, and pruritus were reported with mirabegron use [19].

Patients who are refractory to behavioral and pharmacologic therapy should be properly reevaluated. If the diagnostic criteria of ROAB are fulfilled, a third-line treatment should be offered. Third-line therapy recommended by guidelines is intradetrusor injection of BoNT-A or sacral neuromodulation [6, 14].

3. Mechanism of action of botulinum neurotoxin

Botulinum toxin is a neurotoxin derived from Gram-positive, spore-producing bacteria (*Clostridium botulinum*). The bacteria produce seven serotypes (A–G) of botulinum toxin each with different antigenic profiles and biochemical actions; however, they all have a similar pharmacological effect [20]. Botulinum toxin types A (BoNT-A) and B (BoNT-B) have been developed for clinical use. BoNT-A has the longest duration of action which makes it more suitable for clinical use.

BoNT-A is synthesized as a single polypeptide chain (150 kDa) which is cleaved into a light chain (50 kDa) and a heavy chain (100 kDa) held together by a fragile disulfide bond and noncovalent bonds. The available formulations of BoNT-A are BOTOX (onabotulinumtoxin A) (Allergan, United States), Dysport (abobotulinumtoxin A) (Ipsen, United Kingdom), and Xeomin (incobotulinumtoxin A) (Merz, Germany). Each formulation of BoNT-A has its own dosing regimen which is not interchangeable. BOTOX is most commonly used followed by Dysport (the former is five times more potent than the latter) [20].

Although fibroblast growth factor receptor 3 has been mentioned as a potential BoNT-A receptor, two forms of BoNT-A cell-surface receptors have been identified: gangliosides and the synaptic vesicle-associated protein 2 (SV2) family [21]. BoNT-A's heavy chain attaches to SV2 on the nerve terminals' surface, followed by endocytic internalization of the toxin within the nerve terminal. The toxin is broken within the synaptic vesicle after translocating into the cytoplasm, leaving the light chain of BoNT-A as the actual active moiety. BoNT-A light chain can then cleave synaptosome-associated protein (SNAP25) off the SNARE proteins, a complex protein that when intact forms the core of the neuroexocytosis machinery. This disrupts the fusion of neurotransmitter-containing vesicles with the neuronal cell membrane, inhibiting neurotransmitter release [22].

SV2-immunoreactive and SNAP25-immunoreactive nerve fibers are found in the sub-urothelium and muscle layer of the human bladder, but not in the urothelium. Almost all parasympathetic nerves express SV2 and SNAP25, while only about half of sensory and sympathetic nerves do [23]. Cleaved SNAP25 is the final product of the BoNT-A light chain's enzymatic activity. It is regarded as an appropriate marker of BoNT-A's action and, thus, an essential target in future research [24].

3.1 Motor effect of BoNT-A

Intradetrusor injection of BoNT-A temporarily blocks the presynaptic vesicular release of acetylcholine (ACh) at the neuromuscular junction of the parasympathetic nerves supplying the detrusor muscles and decreases detrusor pressures and phasic contractions in both idiopathic and neuropathic bladders. However, patients also report a significant decrease in urgency and hence, it is hypothesized that botulinum toxin also modulates the sensory pathways [20].

3.2 Sensory effects of BoNT-A

BoNT-A suppresses bladder sensations by processes unrelated to its effects on ACh release. Transient receptor potential vanilloid subfamily 1 (TRPV1) and P2X3 immunoreactive fibers can be seen throughout the sub-urothelium of the human bladder [25]. BoNT-A injection reduces TRPV1 and P2X3 activity in sensory nerve fibers with subsequent reduction in the frequency of urgency episodes [26]. Also, intravesical BoNT-A injections have been demonstrated to reduce ATP and neurotrophin release from urothelial cells and increase NO release [27]. ATP has been shown to play a role in the pathophysiology of OAB by mediating the sense of bladder fullness [27].

4. Techniques of administration of BoNT-A

Intradetrusor injection of BoNT-A should be offered to carefully-selected and thoroughly-counseled patients with ROAB not responding to the previous two lines of treatment. Patients should be able and willing to maintain close follow-up and frequent PVR estimation and accept the possibility of self-catheterization [6].

Patients should be given enough written and verbal peri-procedure instructions, as well as information regarding urinary tract infection (UTI) and urine retention. Prophylactic antibiotics are recommended for all patients and continue for 1–3 days postoperatively. Urinalysis should be performed before the procedure to rule out active infection. Administration of intradetrusor BoNT/A injections has been described in an office setting under local anesthesia or in the operating room with regional or general anesthesia using a flexible or rigid cystoscopy.

For idiopathic detrusor overactivity typically 100–200 units of BOTOX (diluted in 20 mL of normal saline) or 750–1000 units of Dysport have been used. BOTOX 100 U is licensed in Europe to treat OAB with persistent or refractory UUI in adults of both genders [14]. Similarly, the AUA guidelines recommended 100 U of BOTOX as a third-line of treatment of OAB [6].

There is no consensus on the ideal injection technique. The location of injections, the depth of injections, the number of injection sites, and the volume at each site vary in literature. Kuo et al. looked back at injection sites and discovered that success rates were the same whether they were in the bladder body alone, the trigone alone, or the bladder body and trigone together [28]. In a subsequent meta-analysis of OAB patients, no significant differences in efficacy between trigonal sparing and non-trigonal sparing injection techniques were found with short-term cure rates of 52.9% and 56.9%, respectively [29].

The injection depth varies as well and is influenced by the length of the needle tip (available tips range between 2 and 8 mm) used for injections. Some authors added a trace amount of indigo carmine or methylene blue to the injection solution to facilitate observation of the procedure and assessment of drug distribution [30]. Even among the same surgeons, there is likely to be some variation, and no study has determined whether sub-urothelial or intradetrusor injection is preferable. However, intradetrusor injections, as opposed to submucosal injections, with sparing of the trigone are favored. Again, there is no consensus on the number of injection sites and the dilution of the toxin but generally, 20 sites are injected and the volume per injection is usually 0.5–1 mL (**Figure 1**) [31].

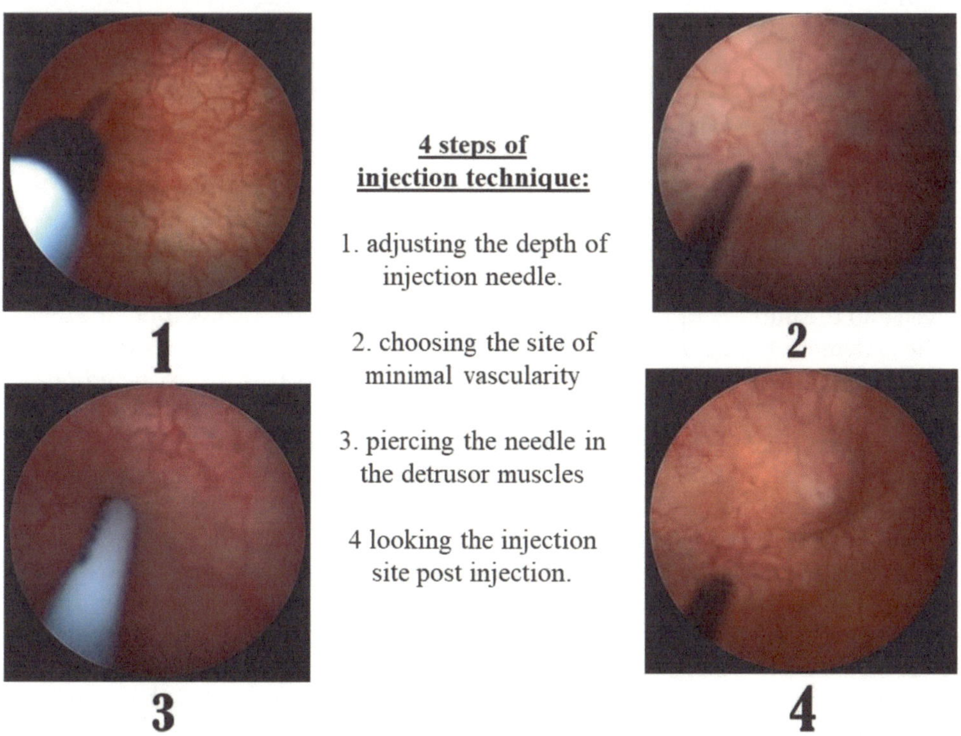

Figure 1.
Illustrate the technique of intra-detrusor injection of BoNT-A.

4.1 Precautions during injection

- Botox is a vacuum-dried protein that must be refrigerated. The vials must be reconstituted with preservative-free saline before injection, and the combination can be kept at 2–8°C for up to 24 h.

- It's worth noting that the product contains human albumin, which should be disclosed due to some patients' reactions to it [31].

- To avoid protein denaturation, avoid rapid shaking of the vial when preparing the combination. Developing an institutional method for labeling syringes containing botulinum toxin dilutions is a significant practical consideration, especially if nurses or other personnel are engaged in its preparation [31].

- Because the thickness of the detrusor muscle varies with the grade of bladder filling, puncturing the muscle appears to be very easy, especially at high bladder filling grades. Furthermore, the bladder wall between trabeculation bars might be quite thin and easily perforated. A suburothelial injection may be a viable option.

5. Efficacy of BoNT-A injection in ROAB

In a multicenter double-blinded randomized trial compared the efficacy of 50, 100, and 150 U onabotulinum toxin A to placebo in OAB patients, at 3 months >50% improvement in urgency and UUI was reported by 37% of the 50 U patients, 68% of 100 U patients and 58% of 150 U. Only the 100 U groups was statistically significant compared to the placebo group. Frequency was statistically significant, reduced in 100 and 150 U groups and complete continence at 3 months was significantly greater in the 100 U group (55%) and the 150 U group (50%) compared to the placebo group (11%). At five-months post-treatment, these differences were maintained [32].

A phase III trial randomized 557 OAB-wet patients; whose symptoms were not responded to anticholinergics to receive bladder wall injections with BoNT-A (100 U) or saline. At week 12, in patients treated with BoNT-A, UUI episodes were halved and the number of micturitions reduced by more than two. A total of 22.9% of the patients in the BoNT-A arm were fully dry, against 6.5% in the saline arm [33]. Tincello and colleagues published the results of the RELAX study including 240 women with refractory OAB to compare 200 U BoNT-A and placebo injections. At 6 months, voiding frequency, urgency, and incontinence episodes per day were significantly reduced in patients receiving BoNT-A injection. Continence was also significantly higher in the treatment group (31% vs. 12%, $p = 0.002$) with subsequent statistically significant improved patients' QOL [34].

In early systematic review assessing the efficacy and safety of botulinum toxin in the management of OAB, Anger et al. reported that BoNT-A treatment improved incontinence episodes and patients' QOL scores, as shown by a 15-point drop in Urinary Distress Inventory scores compared to placebo-injected patients [35]. In a recent systematic review and network meta-analysis, BoNT-A (100 U) had associated with the greatest reductions in urinary incontinence (UI) episodes, urgency episodes, and micturition frequency, and the highest odds of achieving decreases of 100% and ≥50% from baseline in UI episodes/day. In comparison to other pharmacotherapies, BoNT-A was more superior in reduction of UUI episodes, urgency, and frequency and was associated with higher odds of achieving a 100% and ≥50% decrease in UUI episodes/day than most other treatments in the network [36].

Moreover, pre-treatment with BoNT-A improved patients' response to anticholinergic treatment. One hundred pre-treated patients with intravesical injections of 100 IU of BoNT-A and the effect faded, were randomized to receive 10 mg solifenacin and placebo for 12 weeks. After 12 weeks of follow-up, all overactive bladder symptom score items, including the total score, had improved significantly ($P < 0.0001$) in solifenacin group. Also, urodynamic parameters including frequency and amplitude of detrusor contractility and detrusor leak point pressure decreased significantly with increased cystometric capacity and improved incontinence quality of life parameters with solifenacin re-treatment [37]. This was explained by the possibility that repeated BoNT-A injections increased bladder capacity and restored the normal numbers and function of M3 receptors, potentially restoring patients' responsiveness to anticholinergic drugs [37]. This was based on the clinical findings and immunohistochemical assays which evidenced that BoNT-A injections could restore the number and efficacy of intra vesical urothelial and suburothelial receptors [38].

The effects of BoNT-A injection last for 4–10 months (mean 6 months). The median time to request re-treatment in the pooled analysis of the two RCTs was 24 weeks [33, 39]. Follow-up over 3.5 years showed the consistent or increasing duration of effect for each subsequent injection, with a median of 7.5 months [14].

6. Botulinum toxin injection-related adverse events

Intradetrusor injection of BoNT-A is still an invasive procedure and associated with a considerable incidence of adverse events. The local side effects include pain, UTI, bleeding, no benefit, need for further injections, need for temporary self-catheterization. The generalized side effects include flu-like symptoms, dry mouth, and malaise. Bauer et al. focused on side effects with botulinum toxin injection and reported that 54% of patients reported at least one side effect [40]. The side effects included urinary retention (8.9%), gross hematuria (17.9%), UTI (7.1%), dry mouth (19.6%), dysphagia (5.4%), impaired vision (5.4%), eyelid weakness (8.9%), arm weakness (8.9%), and leg weakness (7.1%). However, symptoms other than urinary retention and UTI were transient and resolved without the need for further treatment [40].

The reported rate of UTI ranged from 3.6% to 54.5% from different RCTs [6] and some reported increased rates of UTI with increases in the dose [41].

Increased PVR and the need for self-catheterization is not uncommon adverse event after botulinum toxin injection. The rate of urinary retention in published studies ranges from 5.4 to 43%, depending on how retention is defined [6, 20, 33, 39]. An interim analysis of a long-term extension study found that the proportions of patients requiring CIC remained stable at 4.6%, 4.1%, and 4.7% after one to three BoNT A treatment cycles, respectively [20]. The onset of urinary retention usually coincides with the onset of clinical efficacy, which occurs between 5 and 10 days after injection and the duration of retention varies, with some patients only requiring CIC for a few days, while others require CIC for the duration of the drug's effects [42–44]. Multivariate analysis revealed that preoperative PVR ≥ 100 ml and preoperative bladder capacity were associated with postoperative urinary retention for the first BoNT-A treatment. Preoperative PVR, BoNT-A units injected, and retention after the first injection were all associated with an increased rate of postoperative retention in those who received a second BoNT-A treatment [45].

Mild hematuria is expected to occur transiently after the procedure due to the injection technique, but severe hematuria requiring intervention or hospitalization for bladder irrigation occurs infrequently [44]. Not all patients benefit from treatment, and many patients discontinue injections outside of clinical trials due to a lack of efficacy or intolerable side effects, such as the need for catheterization [43].

7. Botulinum toxin versus SNM for ROAB

The choice between both BoNT-A and SNM for the management of ROAB is influenced by many factors. BoNT-A is a straightforward day case or outpatient procedure, which could be performed with local anesthesia or intravenous sedation. But it is still an invasive procedure. Also, due to the self-limited duration of action, reinjection every 6 months may be necessary. Whereas, SNM is more complex that necessitates a two-step procedure. However, its effect can last for 4–6 years or even longer. Currently, SNM associated risk of injury and surgical complexity are more tolerable due to the standardized and popularity of the technique. As a result, patients may prefer treatment with a longer duration [46]. Safety and effectiveness are critical factors to consider when making a decision. Efficacy and safety of BoNT-A injection and SNM are often assessed using successful treatment rates (symptoms of OAB improvement >50%) and adverse events,

	Sacral neuromodulation (SNM)	**Botulinum toxin (BoNT-A)**
Mechanism of action	SNM mainly functions in the central nervous system: • Somatic afferent nerve activates the centralinhibitory pathway. • Visceral sensory nerve activates the central inhibitory pathway On peripheral nerve: activate the motor nerve pathway	BoNT-A mainly acts on the peripheral nervous system. • BoNT-A can inhibit neurotransmitter release not only from efferent but also possibly from afferent nerve terminals. • Decline the neurotransmitter release in the presynaptic membrane. • Reduces receptors expression in the postsynaptic membrane. • Improves the sensation of the bladder by decreasing ATP in the sub-urothelium. Central effect: • BoNT-A Impairs the sensory fibers terminals in the spinal cord dorsal horn.
Compared to medications	Superior	Equal
Undesirable events	Low incidence	Higher incidence, especially UTI and urine retention
Long term stability	Long term 80% stability	Short term stability 70% of patients drop out
Satisfaction	High maintained satisfaction	Short-term satisfaction
Re-treatment	Less	Repeated in a less than a year
Cost effectiveness	Long term cost effective (5 and 10 years)	Short-term cost effective (2 years)

Table 1.
Comparison of the sacral neuromodulation and botulinum toxin treatment.

respectively. In A multicenter randomized trial that compared the 2-years outcome of BoNT-A and SNM, no difference in mean UUI episodes reduction was found (p = 0.15), with no differences in UUI resolution, ≥75% or ≥50% UUI episodes reduction [46]. Others reported that SNM had a better success rate than BoNT-A after 6 months of follow-up [47]. In a recent systematic review, no significant difference was found in successful treatment between BoNT-A and SNM at 6 months after procedures [48].

The injection of BoNT-A is associated with a significant rate of side effects, particularly urinary retention. Local discomfort and infection are prevalent in SNM and are easily managed. As a result, compared to SNM, BoNT-A injection has a safety disadvantage. Both options have opposing viewpoints on their efficacy.

The suggestion would be reconsidered if we introduced cost-effectiveness. BoNT-A injection would be a cost-effective choice over a two-year period. The results of the ROSETTA randomized trial identified that two-year costs were higher for sacral neuromodulation than for BoNT-A and persisting through 5 years [49]. While at 10 years, SNM provides a considerable possibility of symptom and quality-of-life improvement and is more cost-effective compared to BoNT-A [50]. As a result, in the long run, SNM would be the better alternative.

In the case of the non-responder who was initially treated with SNM or BoNT-A injection, we do not know whether we can switch to another therapy or how effective it will be. BoNT-A can be used in SNM non-responders with a success rate of 43.4% but is associated with a high long-term discontinuation rate (55%) [51].

Furthermore, the success rate in ROAB patients who used SNM therapy after failed BoNT-A therapy was 58.5%. There was no significant difference between ROAB patients who chose SNM as replacement therapy after failed BoNT-A therapy and those who used SNM therapy as first (**Table 1**) [52].

8. Future perspectives

As mentioned before, intra-detrusor injection of BoNT-A is still an invasive procedure that requires anesthesia and is associated with specific complications especially UTI and urine retention. Also, the efficacy and safety of intra-detrusor injection are sensitive to injection volume and depth, and this issue has motivated researchers to study injection-free modes of drug delivery into the bladder [53]. Therefore, intravesical instillation rather than injection of BoNT-A seems to be a sound idea. Nevertheless, BoNT-A delivery to the bladder tissue after intravesical instillation is hampered by toxin degradation by urine proteases, dilution by urine at the time of instillation, and poor uptake due to the urothelium impermeability, which results from the watertight barrier located at the umbrella cells in the superficial layers of bladder urothelium that are augmented by glycosaminoglycan and uroplakins [54].

To overcome this barrier, intravesical instillation of BoNT-A formulated with liposome (lipo-botulinum toxin) to enhance its absorption was evaluated in two studies; one pilot study and a 2-center, double-blind, randomized, placebo-controlled trial. After 1-month, lipo-botulinum toxin instillation was associated with a statistically significant reduced urinary frequency and urgency; however, the treatment did not reduce UUI episodes. Furthermore, onabotulinum toxin complexed with liposomes did not result in urinary retention [55, 56].

Another method tested was to add a chemical agent that enhances drug delivery into the bladder tissue. Dimethyl sulfoxide (DMSO) is an organic solvent that has been used to facilitate the delivery of several anticancer drugs into animal bladders. Petrou et al. studied 25 women with ROAB that were given BoNT-A (300 U) mixed with 50% DMSO. Efficacy and toxicity were assessed at baseline, 1 and 3-months after treatment. The median number of UUI episodes decreased at 1 month ($p = 0.004$) and then increased back at 3 months. Also, a significant reduction in symptom scores from baseline was noted. The Impact Questionnaire short form improved from 13 to 7 at 1 month ($p = 0.007$), and the Urogenital Distress Inventory improved from 10 to 5 at 1 month ($p = 0.003$). No serious side effects or urinary retention were noted [57].

Kodama et al. stated that low energy shock waves (LESWs) increase tissue permeability and drug delivery into cells by the shear force generated by the movement of liquid relative to cells, which temporarily affects the permeability of the plasma membrane. So, it can deliver macromolecular drugs into the cell cytoplasm without toxicity [58]. In the OAB-rat model, intravesical instillation of BoNT-A plus LESW group showed statistically significant lower amplitude.

($p = 0.001$) and lower frequency of detrusor contractions ($p = 0.01$). Histologically, combined treatments significantly reduced submucosal edema and inflammatory cell infiltrate scores. Moreover, BoNT-A plus LESW significantly increased tissue expression of antioxidant marker (superoxide dismutase) and suppressed oxidative stress marker (malondialdehyde) and inflammatory cytokines (tumor necrotic factor-α and interleukin-6) [59].

In preliminary clinical study, including 15 patients with ROAB, Intravesical instillation of 100 IU of BoNT-A was done followed by LESWs (3000 shocks over 10 min) exposure to the suprapubic area was tested. Patients were followed-up by urine analysis, urine culture, PVR, and Overactive Bladder Symptom Score (OABSS) at 1, 2, and 3 months. Patients showed statistically significant improvements in all OABSS domains and the total score after 1 and 2 months of treatment ($P < 0.05$). Whereas, only the nocturia domain remained significantly improved after 3 months ($P = 0.02$). Seven (46.6%) and 12 (80%) patients were totally dry at 1 and 2 months, respectively. Also, treated patients had no significant increase in PVR throughout the study period ($P > 0.05$), and none of the patients required clean intermittent catheterization [60].

Further research to optimize the procedure of injection to be less invasive, more effective, and improve the injection-free mode is mandatory and expected in the near future.

9. Conclusion

- Intravesical injections of BoNT-A have been approved as third-line treatment for OAB after the failure of behavioral and pharmacotherapy with successful short-term outcomes.

- Repeated injections should be put into consideration during decision-making and patient counseling.

- Intravesical BoNT-A injections are associated with a significant rate of adverse events (such as increased post-void residual volume, acute urinary retention, and UTI); thus, informed consent must be given before treatment.

- No consensus on the standard injection technique and dose of BoNT-A in ROAB.

- Approaches to optimize the procedure techniques; to be less invasive, more effective, and with less side effects, improve the injection free mode and improve its outcome to be more durable are mandatory future perspectives.

Conflict of interest

Authors have no conflict of interest to disclose.

Author details

Mohamed H. Zahran[1,2*], Ali Abdel Raheem[3,4], Ibrahim Alowidah[4] and Diaa-Eldin Taha[4,5]

1 Urology Department, Urology and Nephrology Center, Mansoura University, Egypt

2 Urology Department, King Salman Armed Force Hospital, Tabuk, Saudi Arabia

3 Urology Department, Tanta University, Egypt

4 Urology Department, King Saud Medical City, Riyadh, Saudi Arabia

5 Urology Department, Kafr Elsheikh University, Egypt

*Address all correspondence to: zahranmha822@gmail.com; zahranmha@mans.edu.eg

References

[1] Abrams P, Artibani W, Cardozo L, Dmochowski R, van Kerrebroeck P, Sand P. International Continence Society. Reviewing the ICS 2002 terminology report: the ongoing debate. Neurourol Urodyn. 2009;**28**(4):287. DOI: 10.1002/nau.20737

[2] IIrwin DE, Milsom I, Hunskaar S, Reilly K, Kopp Z, Herschorn S et al. Population-based survey of urinary incontinence, overactive bladder, and other lower urinary tract symptoms in five countries: results of the EPIC study. Eur Urol. 2006;**50**(6):1306-1314; discussion 1314-5. DOI: 10.1016/j.eururo.2006.09.019

[3] Zhang Y, Ji F, Liu E, Wen JG. Mechanism and Priority of Botulinum Neurotoxin A versus Sacral Neuromodulation for Refractory Overactive Bladder: A Review. Urol Int. 2021;**105**(11-12):929-934. DOI: 10.1159/000515991

[4] Nambiar A, Lucas M. Chapter 4: Guidelines for the diagnosis and treatment of overactive bladder (OAB) and neurogenic detrusor overactivity (NDO). Neurourol Urodyn. 2014;**33**(Suppl 3):S21-5. DOI: 10.1002/nau.22631

[5] Szymański JK, Słabuszewska-Jóźwiak A, Zaręba K, Jakiel G. Neuromodulation - a therapeutic option for refractory overactive bladder. A recent literature review. Wideochir Inne Tech Maloinwazyjne. 2019;**14**(4):476-485. DOI: 10.5114/wiitm.2019.85352

[6] Gormley EA, Lightner DJ, Faraday M, Vasavada SP. American Urological Association; Society of Urodynamics, Female Pelvic Medicine. Diagnosis and treatment of overactive bladder (non-neurogenic) in adults: AUA/SUFU guideline amendment. J Urol. 2015;**193**(5):1572-80. DOI: 10.1016/j.juro.2015.01.087

[7] Lightner DJ, Gomelsky A, Souter L, Vasavada SP. Diagnosis and Treatment of Overactive Bladder (Non-Neurogenic) in Adults: AUA/SUFU Guideline Amendment 2019. J Urol. 2019;**202**(3):558-563. DOI: 10.1097/JU.0000000000000309

[8] Atiemo HO, Vasavada SP. Evaluation and management of refractory overactive bladder. Curr Urol Rep. 2006;**7**(5):370-375. DOI: 10.1007/s11934-006-0006-0

[9] Raju R, Linder BJ. Evaluation and Treatment of Overactive Bladder in Women. Mayo Clin Proc. 2020;**95**(2):370-377. DOI: 10.1016/j.mayocp.2019.11.024

[10] Banakhar MA, Al-Shaiji TF, Hassouna MM. Pathophysiology of overactive bladder. Int Urogynecol J. 2012;**23**(8):975-982. DOI: 10.1007/s00192-012-1682-6

[11] Andersson KE. Antimuscarinics for treatment of overactive bladder. Lancet Neurol. 2004;**3**(1):46-53. DOI: 10.1016/s1474-4422(03)00622-7

[12] Nambiar AK, Bosch R, Cruz F, Lemack GE, Thiruchelvam N et al. EAU Guidelines on Assessment and Nonsurgical Management of Urinary Incontinence. Eur Urol. 2018;**73**(4):596-609. DOI: 10.1016/j.eururo.2017.12.031

[13] Nitti VW, Rovner ES, Bavendam T. Response to fesoterodine in patients with an overactive bladder and urgency urinary incontinence is independent of the urodynamic

finding of detrusor overactivity. BJU Int. 2010;**105**(9):1268-1275. DOI: 10.1111/j.1464-410X.2009.09037.x

[14] Harding CK, Lapitan MC, Arlandis S, Bø K, Costantini E, Groen J. EAU guidelines on management of non-neurogenic female lower urinary tract symptoms (LUTS). European Association of Urology Guidelines; 2021

[15] Rai BP, Cody JD, Alhasso A, Stewart L. Anticholinergic drugs versus non-drug active therapies for non-neurogenic overactive bladder syndrome in adults. Cochrane Database Syst Rev. 2012;**12**(12):CD003193. DOI: 10.1002/14651858.CD003193

[16] Herbison P, McKenzie JE. Which anticholinergic is best for people with overactive bladders? A network meta-analysis. Neurourol Urodyn. 2019;**38**(2):525-534. DOI: 10.1002/nau.23893

[17] D'Souza AO, Smith MJ, Miller LA, Doyle J, Ariely R. Persistence, adherence, and switch rates among extended-release and immediate-release overactive bladder medications in a regional managed care plan. J Manag Care Pharm. 2008;**14**(3):291-301. DOI: 10.18553/jmcp.2008.14.3.291

[18] Sears CL, Lewis C, Noel K, Albright TS, Fischer JR. Overactive bladder medication adherence when medication is free to patients. J Urol. 2010;**183**(3):1077-1081. DOI: 10.1016/j.juro.2009.11.026

[19] Chapple CR, Kaplan SA, Mitcheson D, Klecka J, Cummings J, Drogendijk T, et al. Randomized double-blind, active-controlled phase 3 study to assess 12-month safety and efficacy of mirabegron, a β(3)-adrenoceptor agonist, in overactive bladder. Eur Urol. 2013;**63**(2):296-305. DOI: 10.1016/j.eururo.2012.10.048

[20] Cruz F. Targets for botulinum toxin in the lower urinary tract. Neurourol Urodyn. 2014;**33**(1):31-38. DOI: 10.1002/nau.22445

[21] Benoit RM, Frey D, Hilbert M, Kevenaar JT, Wieser MM, Stirnimann CU, et al. Structural basis for recognition of synaptic vesicle protein 2C by botulinum neurotoxin A. Nature. 2014;**505**(7481):108-111. DOI: 10.1038/nature12732

[22] Dong M, Yeh F, Tepp WH, Dean C, Johnson EA, Janz R, et al. SV2 is the protein receptor for botulinum neurotoxin A. Science. 2006;**312**(5773):592-596. DOI: 10.1126/science.1123654

[23] Coelho A, Dinis P, Pinto R, Gorgal T, Silva C, Silva A, et al. Distribution of the high-affinity binding site and intracellular target of botulinum toxin type A in the human bladder. Eur Urol. 2010;**57**(5):884-890. DOI: 10.1016/j.eururo.2009.12.022

[24] Coelho A, Cruz F, Cruz CD, Avelino A. Spread of onabotulinumtoxinA after bladder injection. Experimental study using the distribution of cleaved SNAP-25 as the marker of the toxin action. Eur Urol. 2012;**61**(6):1178-1184. DOI: 10.1016/j.eururo.2012.01.046

[25] Schulte-Baukloh H, Priefert J, Knispel HH, Lawrence GW, Miller K, Neuhaus J. Botulinum toxin A detrusor injections reduce postsynaptic muscular M2, M3, P2X2, and P2X3 receptors in children and adolescents who have neurogenic detrusor overactivity: a single-blind study. Urology. 2013;**81**(5):1052-1057. DOI: 10.1016/j.urology.2012.12.021

[26] Apostolidis A, Popat R, Yiangou Y, Cockayne D, Ford AP, Davis JB, et al.

Decreased sensory receptors P2X3 and TRPV1 in suburothelial nerve fibers following intradetrusor injections of botulinum toxin for human detrusor overactivity. J Urol. 2005;**174**(3):977-982; discussion 982-3. DOI: 10.1097/01.ju.0000169481.42259.54

[27] Yiangou Y, Facer P, Ford A, Brady C, Wiseman O, Fowler CJ, et al. Capsaicin receptor VR1 and ATP-gated ion channel P2X3 in human urinary bladder. BJU Int. 2001;**87**(9):774-779. DOI: 10.1046/j.1464-410x.2001.02190.x

[28] Kuo HC. Bladder base/trigone injection is safe and as effective as bladder body injection of onabotulinumtoxinA for idiopathic detrusor overactivity refractory to antimuscarinics. Neurourol Urodyn. 2011;**30**(7):1242-1248. DOI: 10.1002/nau.21054

[29] Davis NF, Burke JP, Redmond EJ, Elamin S, Brady CM, Flood HD. Trigonal versus extratrigonal botulinum toxin-A: a systematic review and meta-analysis of efficacy and adverse events. Int Urogynecol J. 2015;**26**(3):313-319. DOI: 10.1007/s00192-014-2499-2

[30] Szczypior M, Połom W, Markuszewski M, Ciura K, Buszewska-Forajta M, Jacyna J, et al. Overactive bladder treatment: application of methylene blue to improve the injection technique of onabotulinum toxin A. Scand J Urol. 2017;**51**(6):474-478. DOI: 10.1080/21681805.2017.1362467

[31] Cox L, Cameron AP. OnabotulinumtoxinA for the treatment of overactive bladder. Res Rep Urol. 2014;**6**:79-89. DOI: 10.2147/RRU.S43125

[32] Denys P, Le Normand L, Ghout I, Costa P, Chartier-Kastler E, Grise P, et al. Efficacy and safety of low doses of onabotulinumtoxinA for the treatment of refractory idiopathic overactive bladder: a multicentre, double-blind, randomised, placebo-controlled dose-ranging study. Eur Urol. 2012;**61**(3):520-529. DOI: 10.1016/j.eururo.2011.10.028

[33] Nitti VW, Dmochowski R, Herschorn S, Sand P, Thompson C, Nardo C, et al. OnabotulinumtoxinA for the treatment of patients with overactive bladder and urinary incontinence: results of a phase 3, randomized, placebo controlled trial. J Urol. 2013;**189**(6):2186-2193. DOI: 10.1016/j.juro.2012.12.022

[34] Tincello DG, Kenyon S, Abrams KR, Mayne C, Toozs-Hobson P, Taylor D, et al. Botulinum toxin a versus placebo for refractory detrusor overactivity in women: a randomised blinded placebo-controlled trial of 240 women (the RELAX study). Eur Urol. 2012;**62**(3):507-514. DOI: 10.1016/j.eururo.2011.12.056

[35] Anger JT, Weinberg A, Suttorp MJ, Litwin MS, Shekelle PG. Outcomes of intravesical botulinum toxin for idiopathic overactive bladder symptoms: a systematic review of the literature. J Urol. 2010;**183**(6):2258-2264. DOI: 10.1016/j.juro.2010.02.009

[36] Drake MJ, Nitti VW, Ginsberg DA, Brucker BM, Hepp Z, McCool R et al. Comparative assessment of the efficacy of onabotulinumtoxinA and oral therapies (anticholinergics and mirabegron) for overactive bladder: a systematic review and network meta-analysis. BJU Int. 2017;**120**(5):611-622. DOI: 10.1111/bju.13945

[37] Elbaset MA, Taha DE, El-Hefnawy AS, Zahran MH, Shokeir AA. Assessment of Anticholinergic Use After Fading of BTX-A Effects in Refractory Idiopathic Overactive Bladder: A Prospective Blinded Randomized Trial.

Int Neurourol J. 2019;**23**(3):240-248. DOI: 10.5213/inj.1938098.049

[38] Datta SN, Roosen A, Pullen A, Popat R, Rosenbaum TP, Elneil S, et al. Immunohistochemical expression of muscarinic receptors in the urothelium and suburothelium of neurogenic and idiopathic overactive human bladders, and changes with botulinum neurotoxin administration. J Urol. 2010;**184**(6):2578-2585. DOI: 10.1016/j.juro.2010.07.034

[39] Chapple C, Sievert KD, MacDiarmid S, Khullar V, Radziszewski P, Nardo C et al. OnabotulinumtoxinA 100 U significantly improves all idiopathic overactive bladder symptoms and quality of life in patients with overactive bladder and urinary incontinence: a randomised, double-blind, placebo-controlled trial. Eur Urol. 2013;**64**(2):249-256. DOI: 10.1016/j.eururo.2013.04.001

[40] Bauer RM, Gratzke C, Roosen A, Hocaoglu Y, Mayer ME, Buchner A et al. Patient-reported side effects of intradetrusor botulinum toxin type a for idiopathic overactive bladder syndrome. Urol Int. 2011;**86**(1):68-72. DOI: 10.1159/000316080

[41] Dmochowski R, Chapple C, Nitti VW, Chancellor M, Everaert K, Thompson C, et al. Efficacy and safety of onabotulinumtoxinA for idiopathic overactive bladder: a double-blind, placebo controlled, randomized, dose ranging trial. J Urol. 2010;**184**(6):2416-2422. DOI: 10.1016/j.juro.2010.08.021

[42] Visco AG, Brubaker L, Richter HE, Nygaard I, Paraiso MF, Menefee SA et al. Pelvic Floor Disorders Network. Anticholinergic therapy vs. onabotulinumtoxina for urgency urinary incontinence. N Engl J Med. 2012;**367**(19):1803-1813. DOI: 10.1056/NEJMoa1208872

[43] Mohee A, Khan A, Harris N, Eardley I. Long-term outcome of the use of intravesical botulinum toxin for the treatment of overactive bladder (OAB). BJU Int. 2013;**111**(1):106-113. DOI: 10.1111/j.1464-410X.2012.11282.x

[44] Kuo HC, Liao CH, Chung SD. Adverse events of intravesical botulinum toxin a injections for idiopathic detrusor overactivity: risk factors and influence on treatment outcome. Eur Urol. 2010;**58**(6):919-926. DOI: 10.1016/j.eururo.2010.09.007

[45] Osborn DJ, Kaufman MR, Mock S, Guan MJ, Dmochowski RR, Reynolds WS. Urinary retention rates after intravesical onabotulinumtoxinA injection for idiopathic overactive bladder in clinical practice and predictors of this outcome. Neurourol Urodyn. 2015;**34**(7):675-678. DOI: 10.1002/nau.22642

[46] Amundsen CL, Komesu YM, Chermansky C, Gregory WT, Myers DL, Honeycutt EF et al. Pelvic Floor Disorders Network. Two-Year Outcomes of Sacral Neuromodulation Versus OnabotulinumtoxinA for Refractory Urgency Urinary Incontinence: A Randomized Trial. Eur Urol. 2018;**74**(1):66-73. DOI: 10.1016/j.eururo.2018.02.011

[47] Singh R, El Nashar SA, Trabuco EC, Klingele CJ, Gebhart JB, Occhino JA. Comparison of Short Term Outcomes of Sacral Nerve Stimulation and Intradetrusor Injection of OnabotulinumtoxinA (Botox) in Women With Refractory Overactive Bladder. Female Pelvic Med Reconstr Surg. 2015;**21**(6):369-373. DOI: 10.1097/SPV.0000000000000200

[48] He Q, Li B, Zhang C, Zhang J, Luo D, Wang K. Treatment

for refractory overactive bladder: a systematic review and meta-analysis of sacral neuromodulation and onabotulinumtoxinA. Int Urogynecol J. 2021;**32**(3):477-484. DOI: 10.1007/s00192-020-04427-w

[49] Harvie HS, Amundsen CL, Neuwahl SJ, Honeycutt AA, Lukacz ES, Sung VW, et al. Cost-Effectiveness of Sacral Neuromodulation versus OnabotulinumtoxinA for Refractory Urgency Urinary Incontinence: Results of the ROSETTA Randomized Trial. J Urol. 2020;**203**(5):969-977. DOI: 10.1097/JU.0000000000000656

[50] Arlandis S, Castro D, Errando C, Fernández E, Jiménez M, González P, et al. Cost-effectiveness of sacral neuromodulation compared to botulinum neurotoxin a or continued medical management in refractory overactive bladder. Value Health. 2011;**14**(2):219-228. DOI: 10.1016/j.jval.2010.08.006

[51] Baron M, Perrouin-Verbe MA, Lacombe S, Paret F, Le Normand L, Cornu JN. Efficacy and tolerance of botulinum toxin injections after sacral nerve stimulation failure for idiopathic overactive bladder. Neurourol Urodyn. 2020;**39**(3):1012-1019. DOI: 10.1002/nau.24326

[52] Yang G, Xu Y, Qu G, Zhang Y. Refractory overactive bladder patients who chose sacral neuromodulation therapy after failed OnabotulinumtoxinA treatment: A systematic review and meta-analysis. PLoS One. 2020;**15**(3):e0230355. DOI: 10.1371/journal.pone.0230355

[53] Tyagi P, Kashyap M, Yoshimura N, Chancellor M, Chermansky CJ. Past, Present and Future of Chemodenervation with Botulinum Toxin in the Treatment of Overactive Bladder. J Urol. 2017;**197**(4):982-990. DOI: 10.1016/j.juro.2016.11.092

[54] Hsu CC, Chuang YC, Chancellor MB. Intravesical drug delivery for dysfunctional bladder. Int J Urol. 2013;**20**(6):552-562. DOI: 10.1111/iju.12085

[55] Chuang YC, Kaufmann JH, Chancellor DD, Chancellor MB, Kuo HC. Bladder instillation of liposome encapsulated onabotulinumtoxina improves overactive bladder symptoms: a prospective, multicenter, double-blind, randomized trial. J Urol. 2014;**192**(6):1743-1749. DOI: 10.1016/j.juro.2014.07.008

[56] Kuo HC, Liu HT, Chuang YC, Birder LA, Chancellor MB. Pilot study of liposome-encapsulated onabotulinumtoxina for patients with overactive bladder: a single-center study. Eur Urol. 2014;**65**(6):1117-1124. DOI: 10.1016/j.eururo.2014.01.036

[57] Petrou SP, Parker AS, Crook JE, Rogers A, Metz-Kudashick D, Thiel DD. Botulinum a toxin/dimethyl sulfoxide bladder instillations for women with refractory idiopathic detrusor overactivity: a phase 1/2 study. Mayo Clin Proc. 2009;**84**(8):702-706. DOI: 10.1016/S0025-6196(11)60520-X

[58] Kodama T, Doukas AG, Hamblin MR. Shock wave-mediated molecular delivery into cells. Biochim Biophys Acta. 2002;**1542**(1-3):186-194. DOI: 10.1016/s0167-4889(01)00177-x

[59] Nageib M, Zahran MH, El-Hefnawy AS, Barakat N, Awadalla A, Aamer HG, Khater S, et al. Low energy shock wave-delivered intravesical botulinum neurotoxin-A potentiates antioxidant genes and inhibits proinflammatory cytokines in rat model

of overactive bladder. Neurourol Urodyn. 2020;**39**(8):2447-2454. DOI: 10.1002/nau.24511

[60] Nageib M, El-Hefnawy AS, Zahran MH, El-Tabey NA, Sheir KZ, Shokeir AA. Delivery of intravesical botulinum toxin A using low-energy shockwaves in the treatment of overactive bladder: A preliminary clinical study. Arab J Urol. 2019;**17**(3):216-220. DOI: 10.1080/2090598X.2019.1605676

Chapter 5

Gummy Smile and Treatment with Botulinum Toxin Type A (Botox)

Imad Katbeh, Mohammad Osama Makkeiah, Tamara Kosyreva and Lada Saneeva

Abstract

A smile plays an important role in determining a person's initial impression, and its assessment has become integral to clinical evaluation. A smile with an esthetic appearance should be symmetrical and should reveal less than 2 mm of gums when smiling. A gingival smile, gummy smile, or high smile line, is defined as the number of excess gums on the upper jaw exposed. This may have some serious psychological repercussions on the patient, which may sometimes lead them to conceal their smile to avoid "embarrassment." One of the most common methods of treating a gingival smile resulting from an overactive lip is lip reposition as a surgical procedure and the injection of type A (Botox) toxin as an injectable inhibitor of muscle action. However, many patients refrain from surgical treatment because of fear of complications and pain. In this case, injections of botulinum toxin group A are an excellent alternative to surgery. The injection of botulinum toxin takes less time and with the correct dosage and compliance with the protocol of its administration causes much fewer complications. The study presented here is devoted to the disclosure of the potential of this tool in esthetic dentistry.

Keywords: lip repositioning, gummy smile, hyperactive upper lip, excessive gingival display, botulinum toxin type A

1. Introduction

Since facial expression and smile are the main components of nonverbal communication, as they play an important role in creating an initial favorable impression in business and friendly communication; therefore, they are given such importance when conducting a clinical evaluation. It should be borne in mind that when the facial expression changes, it can show pleasure, sympathy, indifference, approval or disgust, amazement or fright, thus various facial defects, including a gummy smile (excessive gingival display), can make it difficult for the patient to build relationships with other people, and if he/she is suspicious, deprive him/her of psychological comfort and contribute to the emergence of various psychological complexes.

In modern society, great importance is attached to the esthetics of appearance, this is what greatly contributed to the development of the elite direction of esthetic dentistry and cosmetology, the patient's smile has become an important marker of their viability and activity, not to mention its influence of this parameter on the patient's self-confidence, which we mentioned earlier. Various structures of the face and the dentoalveolar system are involved in the formation of a smile, respectively, any defect in them will cause a change in the smile and, most likely, will require correction, that is, certain medical procedures.

A smile that meets high esthetic standards should be symmetrical, it is believed that with a beautiful smile, less than 2 mm of gums should be exposed, not every patient meets this standard, so recently, it has become common for many people to hide their smile when they are happy or photographed. This phenomenon is known as a "gummy smile," which is a large exposure of the gums when smiling, creating an appearance far from pleasant or desirable in the eyes of others.

2. Gummy smile

This excessive exposure of the gums in a full smile occurs as a result of an imbalance in the interaction of the three main components modeling the smile mentioned above—lips, gingiva, and teeth.

Hulsey pointed out that the most attractive smiles were those in which the upper lip reached the level of the gingival edge of upper incisors [1].

Chiche and Pinault concluded that exposure of up to 3 mm of the gingival tissue could still be considered satisfactory [2]. Hunt et al. argued that the acceptable range of gingival exposure was between 0 and 2 mm, with the ideal situation being no exposure at all [3]. Geron and Atalia found that any maxillary gingival exposure above 1 mm would be considered unesthetic [4].

Since other researchers have recently discovered that 3 mm of gum exposure worsen the appearance of the patient, which does not meet the esthetic expectations of modern society, thus 2 mm of gum exposure would be the maximum for an esthetically acceptable smile, and therefore, it can be said that the ideal amount of gum exposure with a full smile is 0–2 mm. In other words, an exposure of up to 3 mm can still be considered esthetically acceptable, but if it is over 3 mm with a full smile, then such a smile should be called a "Gummy smile" or "Gingival smile" [4–6].

The prevalence of a gummy smile is 10–15% at the age of 20–30 years, but with age, due to sagging muscle tissue of the upper lip and its covering with gingival tissue along with part of the teeth of the upper jaw, the incidence decreases. The prevalence of gummy smile varies according to gender. A study found that the amount of gingiva exposed when smiling in females was 1.5–2 mm greater than males, at a rate of 7% in males and 14% in females [7].

3. Treatment of gummy smile caused by dentoalveolar factors

3.1 Treatment of gummy smile caused by shortening the clinical crown length of maxillary anterior teeth

Here, the distinction should be made between the two cases.

If shortness in the clinical length is caused by the gums covering the crowns of maxillary anterior teeth, it is treated with a traditional gingivectomy [8].

If there is a shortening in the clinical crown length of maxillary anterior teeth with normal gingival growth, here it is necessary to remove an amount of the bone as well as treat this case by surgically lengthening the teeth.

One of the criteria that should be followed before starting to lengthen the crowns is the availability of a sufficient amount of attached gingiva of no less than 3 mm to reduce postoperative gingival recession and maintain vital presentation [9].

3.2 Treatment of gummy smile caused by dentoalveolar eruption

Treatment is through orthodontic work, correction of the axis of maxillary anterior teeth, and orthodontic intrusion. Sometimes, surgical intervention might be necessary in cases of deep bite and excessive overjet following a comprehensive clinical and radiographic study.

In other words, bone, periodontal ligament, and soft tissues move with teeth movement, so intrusion helps to improve gummy smiles. On the other hand, with the emergence of orthodontic implants, which are known as temporary anchorage devices, anterior teeth intrusion has become a possibility.

3.3 Treatment of gummy smile caused by gingival hyperplasia

Gingival hyperplasia accompanied by periodontal pockets is treated with conventional surgical resection of the gum, necessary to achieve optimal periodontal condition and reduce the depth of periodontal pockets so that they do not exceed the mucosal-gingival junction.

3.4 Treatment of gummy smile caused by soft tissue factors

Shortening of the upper lip and hyperactivity of levator labii superioris muscle.

Traditional orthodontic surgical methods of treating gummy smile are perceived by patients to be painful and unacceptable methods. Therefore, it was necessary to devise less invasive and more patient-acceptable methods to fulfill esthetic demands. These methods include surgical lip repositioning, injection of botulinum toxin type A (Botox), and maxillary anterior nasal spine implants.

3.4.1 Surgical repositioning of the lip

This is a surgical technique performed under local anesthesia to reposition the lip, which helps to reduce gingival exposure by limiting muscle tension of the levator labii superioris muscle by decreasing the depth of the vestibule. This is achieved through the removal of a spindle-shaped strip of the vestibular fold either with cutting insertions of the levator labii superioris muscle (fully thick slice) [10] or without (partly thick slice) [11].

Indications for surgical repositioning of the lip:

1. Gummy smile caused by shortness of the upper lip or hyperactivity of levator labii superioris muscle.

2. Good oral health and safety of periodontal tissues.

3. Stable systemic health.

4. Sufficient amount of attached gingiva no less than 3 mm.

Contraindications for surgical lip repositioning:

1. Gummy smile caused by structural factors.
2. The amount of attached gingiva is less than 3 mm.
3. Local or systemic contraindications of periodontal surgical procedures.
4. Pregnant and breastfeeding women.
5. Smoking.

3.4.2 Anterior nasal spine implants

Maxillary anterior nasal spine implantation obstructs lifting of the upper lip, which helps to reduce gingival exposure when smiling.

Austin describes this technique; a pocket is created by bilaterally incising and raising the periosteum from the maxillary anterior nasal spine by 10 mm. The pocket is then filled with silicone and the latter is left to solidify and take on the shape of the area. It is then removed, the hard edges trimmed, re-implantation performed, and the pocket sutured. Results were good and patients were fully satisfied. However, during such interventions, infection of the periodontal pocket may occur, and in this case, it will be necessary to remove all its contents and administer antibiotics [12].

3.4.3 Botox injection

3.4.3.1 Botox: concept and history

It is considered one of the non-surgical alternatives to reduce gummy smile caused by the hyperactivity of the levator labii superioris muscle. This technique has become a part of clinical practice [13].

The term Botox is a combination of two words—botulinum and toxin, where the first refers to the name of the microbes from which this substance is extracted, and the second means poison in English. Thus, it consists of substances extracted from the bacterium clostridium botulinum [14, 15].

The Botox known to us today was developed by US ophthalmologist Alan Scott in 1970, who said the following about his product: "When I developed it, I knew that it could do miracles with regard to neurological problems, but I never knew that it might work as a cosmetic substance" [16].

At first, Alan Scott and his colleague Schantz used low doses of serotype BTX A to treat cases of "strabismus" and cases of blepharospasm. Thus, initially, since the 1970s, this drug was used to treat strabismus and then found its application in cosmetology and esthetic medicine.

In 1989, FDA approved Allergan's BTX A as an effective and safe treatment for blepharospasm, strabismus, hemifacial spasm, and 7th cranial nerve disorders. In April 2002, it obtained approval to treat glabella lines associated with corrugator supercilii and procerus muscles activity. In 2004, it was approved as a treatment for axillary hyperhidrosis [17].

3.4.3.2 Biochemical and physical properties of Botox

Clostridium botulinum, which is a gram-positive, anaerobic spore-forming bacterium, produces multiple exotoxins.

The most stable and potent exotoxin is BTX A. Its median lethal dose in monkeys is 39 U/kg—approximately 2500–3000 U/kg in humans.

BTX A has a neurotoxic effect by suppressing synaptic transmission in cholinergic nerve endings. When it overlaps with presynaptic terminations, it prevents the extracellular release of acetylcholine, that is, it can be used to reduce skeletal muscle tone by binding to a synaptosomal protein (SNAP-25), which inhibits the release of acetylcholine from motor neurons and activates presynaptic repolarization [18]. Side effects are limited to the peripheral nervous system and mainly occur through the neuromuscular junction (NMJ), but it should be noted that the nerve endings of the ganglia can also be affected.

When it affects NMJ's nerve endings, BTX A causes denervation of muscle fibers and triggers reversible paralysis. It should be noted that denervated muscles can atrophy after BTX A administration, but they restore sensitivity to neurotransmission by producing acetylcholine receptors outside the compound. Motor neurons also develop new endings, so if given enough time to recover, they will eventually reverse the paralysis. When BTX A is administered for therapeutic purposes, its effectiveness is limited to a few months [19].

3.4.3.3 Therapeutic applications

The number of therapeutic applications of BTX A has significantly expanded after years of study on humans, for whom local injection of BTX A is the primary treatment for blepharospasm, spastic dysphonia in addition to strabismus and hypersalivation, myoclonus, stuttering, nystagmus, tremor, and focal hyperhidrosis.

3.4.3.4 Botox cosmetic indications

1. Cosmetic procedures to eliminate wrinkles, which account for about 90% of all cosmetic applications of Botox.

2. Elimination of glabellar wrinkles; deep glabellar wrinkles occur due to constant contractions of the glabella muscles arising from various causes.

3. Periorbital wrinkles, for the elimination of which Botox is one of the most suitable means, most often they are formed in old age, due to constant contractions of the circular muscle of the eye.

4. Wrinkles on the forehead; these wrinkles develop with age and due to the habit of raising eyebrows when expressing surprise, patients often seek to get rid of them, as they form an appearance that does not meet their esthetic needs.

5. Correction of the shape of the nose: Botox is used for contour non-surgical rhinoplasty, since when it is administered, the muscles that pull the tip of the nose down are weakened, which allows you to achieve the desired shape of the nose for the patient, due to the action of antagonist muscles.

6. It is used to treat age-related wrinkles on the neck or the so-called "chicken neck."

7. Botox is also used to treat cervical dystonia, which is a spasm in the neck muscles, in which spasms may occur in the muscles of the hands.

8. It is used for dermal diseases, especially hyperhidrosis in armpits, where it is intradermally injected.

9. Correction of gingival smile resulting from hyperactive upper lip, BTX-A injection exhibits better results than those of surgery and has given safer and more satisfactory cosmetic outcomes [20].

10. Botox can be used to improve the shape of the wound after suturing, it prevents muscle tension at the edges of the wound and promotes the formation of a thinner cosmetic scar [14, 21].

Botox treatment can be recommended to a patient of any age since the appearance of wrinkles is usually associated with high activity of facial muscles, and not with age-related skin changes in patients. Therefore, it is possible to come across patients who are in their twenties. It is worth noting that its result will be much better if Botox is used before the formation of deep wrinkles, which are difficult to remove even during the facelift. The complexity of the procedure depends on the severity, depth and location of wrinkles, and primarily on the quality and thickness of the skin. It should be noted that the use of Botox is limited in the area of the circular muscle of the mouth due to its active participation in the processes of eating and speech formation.

3.4.3.5 Contraindications

Absolute contraindications:
Botox is not used in the following cases:

1. During pregnancy, when intending to become pregnant as well as breastfeeding.

2. In the event of a severe allergy to Botox, which represents a rather small proportion of cases.

3. In case of inflammation or swelling in the area planned for Botox injection.

4. If the patient has a neuromuscular disease.

5. Use of alcohol and aspirin before treatment, it can be carried out no earlier than two weeks after the last intake of these drugs.

Relative contraindications:

1. Amyotrophic lateral sclerosis.

2. Myasthenia gravis.

3. Lambert-Easton syndrome.

4. Respiratory disorders, such as asthma and Bloating.

5. Swallowing problems.

6. Weakness in facial muscles (drooping eyelid, or inability to raise eyebrows).

7. Hemorrhagic and heart problems.

8. Previous Botox administration was carried out less than 4 months ago.

9. Occurrence of various complications with the introduction of Botox in the anamnesis.

Disadvantages of Botox

1. Allergic reactions to this drug, which are quite rare.

2. Incomplete clinical effect from the use of the drug or even asymmetry between the right and left sides of the face; which can be corrected by repeated administration of the drug in the areas indicated 15 days after the first administration.

3. Occurrence of twitching in the muscles located near the site of administration of the drug, lasting for several hours, which subsequently disappears.

4. Ptosis of the upper eyelid, which occurs due to exceeding the dosage of the drug or non-compliance with the drug administration algorithm, according to which the injection site for the correction of frontal wrinkles should be 2 cm above the upper edge of the orbit; this condition disappears by itself two to three weeks after administration of the drug.

5. In rare cases, corneal dryness may occur, lasting for a short period of time, amenable to correction with the help of eye moisturizers.

6. Numbness or burning sensation in the area of administration of the drug, lasting for a short period of time after its administration.

7. Swelling or hyperemia that occurs after administration of the drug, usually passing within an hour after its administration.

Botox is used in dentistry and maxillofacial surgery in the following cases:

1. Pathological TMJ disorders,

2. Bruxism,

3. Masticatory muscle tension disorders,

4. Mandibular muscle spasms,

5. In the treatment of gummy smiles.

3.4.3.6 Gummy smile and treatment with type A botulinum toxin injection

In a pioneering study by Polo [22], botulinum toxin injection to treat gummy smile showed promising results, as it did not require either surgery or anesthesia. Instead, a high cosmetic effect was achieved by reducing the amount of gum exposure by inhibiting and partially blocking by introducing Botox type A into the projection area of the following muscles—the levator labii superioris and the zygomatic muscle [22].

In another study by Polo [23] that involved 30 patients, two points were injected with five units of Botox bilaterally above the levator labii superioris alaeque nasi and zygomaticus minor muscles. Gingival exposure improved by 0.09 mm two weeks after injection.

Hwang [16] performed measurements on corpses to determine the Yonsei point for a single injection of targeted muscles, including the levator labii superioris alaeque nasi. The Yonsei point was determined to be at the center of a triangle formed by the intersection of the muscle's levator labii superioris, levator labii superioris alaeque nasi, and zygomaticus minor muscles, laterally located 1 cm on the transverse plane, and 3 cm above the lip line on the frontal plane in both men and women.

In a 2010 study, Mazzuco and Hexcel classified patients with a gummy smile into three categories:

Anterior gummy smile—a single injection point in the lateral nasolabial fold 1 cm below the levator labii superioris alaeque nasi.

Posterior gingival smile: two injection points—the first in the nasolabial fold at the point of maximum contraction when smiling; the second, 2 cm lateral to the first point at the level of the tragus of the ear (zygomaticus minor and major).

Mixed anterior and posterior gingival smile—both anterior and posterior injection points (levator labii superioris alaeque nasi and zygomaticus minor and major muscles).

Asymmetrical gingival smile—two posterior injection points on the side of the greater gingival exposure and only in the lowest point on the opposite side (zygomaticus minor and major) [15].

Sucupira and Abramovitz [24] used only one injection point 3–5 mm lateral to the nostril, targeting the levator labii superioris alaeque nasi muscle. They reported 84% improvement rate.

As for Suber, he used three injection points: 2 mm lateral to the pterygopalatine fossa at the level of the nasal passage, followed by another 2 mm lateral to the first at the same transverse level, and a third injection 2 mm below and between the first and second sites. The resulting injection sites were drawn as an inverted triangle [25].

Dinker et al. used Botox type A to manage the problem of gummy smile [26].

Aly and Hammouda [27] proposed combining Botox injection and lip repositioning for cases of gummy smile resulting from vertical maxillary excess.

Similarly, Pedron and Mangano [28] discussed the benefit of combining gingivectomy and Botox injection in the management of cases of the gingival smile. They used graded 1 mL 30 G insulin syringes, and, as for type A botulinum toxin, a 100-U vial, along with sodium chloride solution (serum), cotton, and alcohol.

3.4.3.6.1 Type A botulinum toxin injection technique

Injection sites are to be determined in three places on each side through muscular activity (smiling) and probing the extent of contraction to ensure accurate muscle position before injection, given that there might occur minor anatomical site variations with no local anesthesia performed.

Surface facial marks used for locating injection sites according to Suber [25]. The resulting injection sites were drawn as an inverted triangle. The bone marker was the forefront of the maxilla attached to the muscles covering it—levator labii superioris alaeque nasi and zygomaticus minor [25].

The amount of botulinum toxin used for each target muscle was 2.5 units. A 100-unit bottle of dried botulinum toxin type A dissolved by adding 2 mL chloride saline serum under sterile conditions so that the dilution ratio was ½, as each grade on the syringe indicated 1 unit of diluted and injectable Botox.

About 1 mL insulin syringes with 30 G needles to be used for injection [29]. Grover et al. confirmed that 1 mL insulin syringes with 26–30 G needles are preferred for Botox injection since the bevel is much thinner, sharper, and long enough to enable injection in facial muscles. Additionally, the mL-graded syringe makes unit distribution easier [30].

After identifying the three injection sites and without performing local anesthesia, the area to be wiped with cotton moistened with alcohol for cleansing, and then 2.5 units should be injected frontally and bilaterally into each of the three muscles, according to Polo and Sumaya [31].

4. Conclusion

Carrying out injections of botulinum toxin type A for the correction of the gingival smile is a promising direction of esthetic dentistry. Of course, the method is not without individual drawbacks—partial relapses of this condition, requiring repeated administration, edema, the occurrence of compensatory wrinkles, and compensatory facial expressions or allergic reactions. Nevertheless, this method should be recognized as less traumatic compared to surgical intervention, which also has several complications and does not guarantee the absence of relapses of this pathology.

In addition, this method is usually recommended for patients who need correction of the gingival smile for esthetic purposes, who refuse surgical intervention for psychological reasons (fear of the sight of their own blood, the occurrence of pain, fear of notwithstanding a long surgical operation).

We believe that the development of esthetic dentistry will follow the path of developing minimally invasive interventions that achieve a great therapeutic effect with simple techniques such as the implementation of botulinum toxin type A injection.

Conflict of interest

The authors declare no conflict of interest.

Author details

Imad Katbeh[1*], Mohammad Osama Makkeiah[2], Tamara Kosyreva[1] and Lada Saneeva[1]

1 Peoples' Friendship University of Russia (RUDN University), Medical Institute, Moscow, Russia

2 Damascus University, Damascus, Syria

*Address all correspondence to: katbeh@bk.ru

References

[1] Hulsey CM. An esthetic evaluation of lip-teeth relationships presents in the smile. American Journal of Orthodontics and Dentofacial Orthopedics. 1970;**57**(2): 132-144

[2] Chiche GJ, Pinault A. Smile rejuvenation: A methodic approach. Practical Periodontics and Aesthetic Dentistry. 1993;**5**(3):37-44. quiz 44. 38

[3] Hunt O, Johnston C, Hepper P, Burden D, Stevenson M. The influence of maxillary gingival exposure on dental attractiveness ratings. European Journal of Orthodontics. 2002;**24**(2):199-204

[4] Geron S, Atalia W. Influence of sex on the perception of oral and smile esthetics with different gingival display and incisal plane inclination. The Angle Orthodontist. 2005;**75**(5):778-784

[5] Bynum J. Treatment of a "gummy smile": Understanding Etiology is key to success. Compendium of Continuing Education in Dentistry (Jamesburg, N.J.: 1995). 2016;**37**(2):114-122

[6] Ezquerra F, Berrazueta MJ, Ruiz-Capillas A, Arregui JS. New approach to the gummy smile. Plastic and Reconstructive Surgery. 1999;**104**: 1143-1150

[7] Zardawi FM, Gul SS, Fatih MT, Hama BJ. Surgical procedures reducing excessive gingival display in gummy smile patients with various etiologic backgrounds. Clinical Advances in Periodontics. 2020;**10**(3):130-134

[8] Trushkowsky R, Montalvo Arias D, David S. Digital smile design concept delineates the final potential result of crown lengthening and porcelain veneers to correct a gummy smile Int. Journal of Esthetic Dentistry. 2016;**11**(3):338-354

[9] Mahn DH. Elimination of a "gummy smile" with crown lengthening and lip repositioning. The Compendium of Continuing Education in Dentistry. 2016;**37**(1):52-55

[10] Abdullah WA, Khalil HS, Alhindi MM, Marzook H. Modifying gummy smile: A minimally invasive approach. The Journal of Contemporary Dental Practice. 2014;**15**:821-826

[11] Simon Z, Rosemblatt A, Dorfmann W. Eliminating a gummy smile with surgical lip repositioning. Journal of Cosmetic Dentistry. 2007;**23**:100-108

[12] Austin HW. Correction of the gummy smile--a plastic surgeon's view. Dentistry Today. 1990;**9**(2):28

[13] Araujo JP, Cruz J, Oliveira JX, Canto AM. Botulinum toxin type-a as an alternative treatment for gummy smile: A case report. Dermatology Online Journal. 2018;**24**(7):13030/qt75f0h8kz

[14] De Maio M, Rzany B. Botulinum toxin in aesthetic medicine. Berlin, Heidelberg: Springer; 2007. 141 p. DOI: 10.1007/978-3-540-34095-9

[15] Mazzuco R, Hexsel D. Gummy smile and botulinum toxin: A new approach based on the gingival exposure area. Journal of the American Academy of Dermatology. 2010;**63**(6):1042-1051

[16] Hwang W-S, Hur M-S, Hu K-S, Song W-C, Koh K-S, Baik H-S, et al. Surface anatomy of the lip elevator muscles for the treatment of gummy smile using botulinum toxin. The Angle Orthodontist. 2009;**79**(1):70-77

[17] Chen S. Clinical uses of botulinum neurotoxins: current indications,

limitations and future developments. Toxins (Basel). 2012 Oct;**4**(10):913-939. DOI: 10.3390/toxins4100913. Epub 2012 Oct 19. PMID: 23162705; PMCID: PMC3496996

[18] Meunier FA, Schiavo G, Molgó J. Botulinum neurotoxins: From paralysis to recovery of functional neuromuscular transmission. Journal of Physiology Paris. 2002 Jan-Mar;**96**(1-2):105-113. DOI: 10.1016/s0928-4257(01)00086-9. PMID: 11755789

[19] Aoki K. Evidence for Antinociceptive activity of botulinum toxin type A in pain management. Headeach. 2003;**43**(suppl. 1):S9-S15

[20] Makkeiah MO, Harfoush M, Makkiah A, Saneeva L, Tuturov N, Katbeh I. Comparative efficacy of Botox and surgical lip repositioning in the correction of gummy smile. Dentistry Stomatologiia. 2021;**100**(3):47-54. (In Russ.). DOI: 10.17116/stomat202110003147

[21] Jaspers GWC, Pijpe J, Jansma J. The use of botulinum toxin type A in cosmetic facial procedures. International Journal of Oral and Maxillofacial Surgery. 2011;**40**:127-133

[22] Polo M. Botulinum toxin type A in the treatment of excessive gingival display. American Journal of Orthodontics and Dentofacial Orthopedics. 2005;**127**(2):214-218 quiz 261

[23] Polo M. Commentary on: Botulinum toxin for the treatment of excessive gingival display: A systematic review. Aesthetic Surgery Journal. 2016;**36**(1):89-92

[24] Sucupira E, Abramovitz A. A simplified method for smile enhancement: Botulinum toxin injection for gummy smile. Plastic and Reconstructive Surgery. 2012;**130**(3): 726-728

[25] Suber JS, Dinh TP, Prince MD, Smith PD. Onabotulinumtoxin A for the treatment of a "gummy smile". Aesthetic Surgery Journal. 2014;**34**(3):432-437

[26] Dinker S, Anitha A. Management of gummy smile with botulinum toxin type-A: A case report. Journal of International Oral Health. 2014;**6**(1): 111-115

[27] Aly LA, Hammouda NI. Botox as an adjunct to lip repositioning for the management of excessive gingival display in the presence of hypermobility of upper lip and vertical maxillary excess. Dental Research Journal (Isfahan). 2016;**13**(6):478-483

[28] Pedron IG. Comment on "botulinum toxin type-A as an alternative treatment for gummy smile: A case report". Dermatology Online Journal. 2019;**25**(6):13030/qt1qk3183b

[29] Amin V, Swathi, Jabir, Shetty P. Enhancing the smile with Botox – Case report. Global Journal of Medical Research. 2013;**13**(2):14-18

[30] Grover S, Malik V, Kaushik A, Divakar R, Vadav P. A future perspective of botox in dentofacial region. Journal of Pharmaceutical and Biomedical Research. 2014;**04**(5):525-531

[31] Polo M. Botulinum toxin type a (Botox) for the neuromuscular correction of excessive gingival display on smiling (gummy smile). American Journal of Orthodontics and Dentofacial Orthopedics. 2008;**133**(2):195-203

Chapter 6

Quality of Life in Neurogenic Bladder Patients and Improvement after Botulinum Toxin Injection

Giovanni Palleschi and Antonio Cardi

Abstract

Various neurogenic conditions may determine an altered function of the bladder and urethral sphincters leading to urinary symptoms. Among these symptoms, loss of urine is considered the most bothersome, and recent literature has proposed that urinary incontinence and poor quality of life should be considered as associated conditions. Urinary incontinence is responsible for reducing the enjoyment of life, including loss of self-confidence and limitation of social activity; it is also associated with anxiety, depression, and deterioration in sexual life, and reduced physical activity. Conservative treatments represent a first-line therapeutic approach to neurogenic urinary disorders, followed by oral medications. However, these treatments often do not provide complete recovery from symptoms, especially from urinary incontinence. Onabotulimun toxin A has proven to be safe and effective for treating neurogenic urinary incontinence and its use is associated with a strong improvement of patients' quality of life. Furthermore, this treatment lowers the risk of severe complications to the upper urinary tract, reducing the need for hospitalization. Quality of life improvement as well as clinical efficacy must be considered targets of the therapy, and the use of onabotulimun toxin A for treating neurogenic urinary incontinence reaches both these goals.

Keywords: neurogenic urinary incontinence, quality of life, onabotulinum toxin A

1. Introduction

Various neurogenic conditions are responsible for lower urinary tract symptoms (LUTS) which further worsen the quality of life (QoL) of patients. In most cases, LUTS onset follows a neurogenic diagnosis by several years, while in certain disorders (i.e., parkinsonism) the urologic disease can sometimes precede neurogenic symptoms. However, in all cases, LUTS represent a significant problem for neurogenic patients, especially for those suffering from physical limitations, such as spinal cord injured (SCI) ones. In addition, LUTS of neurogenic origin is often secondary to a bladder-sphincteric dysfunction which can lead to severe complications and, therefore, must be early and properly managed. The aim of treatment is to restore adequate bladder filling and emptying processes, preventing severe complications, especially to the upper urinary tract (renal failure), and improving QoL. Various therapeutic options are today available for neurogenic LUTS, varying from conservative approaches (behavioral treatment, physical therapy) to pharmacologic oral medications (antimuscarinics, beta-3- agonists, alpha-blockers). Among LUTS, urinary incontinence (UI) is the most bothersome and, in the neurogenic bladder (NGB), is often severe because it is the consequence of neurogenic detrusor overactivity (NDO), which is characterized by involuntary bladder contractions. Conservative treatments and first-line therapeutic options often fail to provide complete recovery from neurogenic UI. In these cases, the international guidelines therapeutic algorithms suggest the use of mini-invasive treatments, including onabotulinum toxin A (BoTA) which has proven to be effective in restoring urinary continence, improving urodynamic parameters, and ameliorating QoL. This chapter aims to report the current knowledge about the relationship between neurogenic disorders and LUTS, their impact on patients' QoL, and how they improve after administration of BoTA, as shown by large cohort studies provided by literature.

1.1 Neurogenic bladder: Definition and epidemiologic data

Neurogenic bladder refers to bladder dysfunction secondary to neurologic disease affecting any point of the complex neuronal circuit, which can ultimately compromise safe bladder filling and emptying [1]. Various pathologic conditions of the central and peripheral nervous system can lead to altered bladder and urethral sphincter function, and consequently, many people suffering from the neurogenic disease can develop LUTS. Epidemiological data available show that, in the United States, NGB has been found in 40–90% of patients with multiple sclerosis (MS), in 37–72% of patients with parkinsonism, and 15% of patients with stroke [2, 3]. In the same country, it is estimated that 70–84% of patients with SCI have at least some degree of bladder dysfunction, which is also frequently seen in patients with spina bifida, associated with vesicoureteral reflux in 40% of children and UI in 60.9% of young adults [4]. Less common scenarios for NGB may include diabetes mellitus (due to autonomic neuropathy involving the bladder), unintended sequelae following pelvic surgery, and cauda equina syndrome resulting from lumbar spine pathology [5]. More specific epidemiological data are provided by studies on neurogenic UI. A meta-analysis of five studies showed that men who suffered a stroke were at increased risk for UI with a pooled odds ratio 0f 2.68 [6]. Other reports showed that men who had a stroke presented an increased risk for UI with an odds ratio ranging from 7.1 to 8.26 [7, 8]. These studies reported that among UI sufferers, the stroke survivors had a higher prevalence of UI compared to controls (17% vs. 9%). Poorer data are provided by Literature regarding neurogenic UI in women. However, various studies analyzed the association between UI and dementia in females. Even if data in some cases are controversial, in a 9-year follow-up of 1453 women aged 65 years and over enrolled in a US HMO, diagnosed dementia was strongly associated with an incident diagnosis of UI (odds ratio of 3.0, ranging from 2.4 to 3.7, 95% CI) [9]. Given the strength and consistency of association with prevalent and incident UI, and given that treatment for reversible dementia can improve UI, a causal role seems certain [9, 10]. Among NGB patients are those suffering from the non-neurogenic neurogenic bladder (the Hinman syndrome), a pathological condition characterized by the uncoordinated activity of the lower urinary tract muscles. In these patients the contraction of the sphincter during voiding and overactivity of the detrusor muscle may lead to urinary frequency and incontinence, reproducing a neurogenic voiding pattern that can determine complications to the bladder wall like those secondary to NGB [11].

NGB is responsible for symptoms of both the bladder filling and voiding phase. Symptom association and severity are strongly related to the site of the neural system involved by the pathophysiological mechanisms of the specific neurogenic condition. Therefore, it is important for clinicians, caregivers, and patients, to gain adequate knowledge of bladder neurophysiology and pathophysiology of NGB.

1.2 Neural control of the bladder and pathophysiology of neurogenic bladder

Bladder dysfunction secondary to a neurogenic disease can be classified considering the location of the neurologic lesion in the nervous system [12]. This type of classification has the most clinical utility to manage patients with NGB from diagnosis to treatment. Neural control of the bladder and urethral sphincters is extremely complex. However, the different areas of the central and peripheral nervous system play specific and highly defined roles during the bladder filling and emptying. Various studies, including urodynamic investigations, neurophysiologic tests, and advanced imaging techniques (PET, functional MRI) allowed the precise identification of the parts of the nervous system that are activated during the different phases of the micturition cycle. The storage phase of the micturition cycle (bladder filling) is maintained by inhibition of parasympathetic activity, and consequent active relaxation of the bladder mediated by the sympathetic system acting on beta 3 adrenoceptors of the detrusor muscle. During the bladder filling, the sympathetic and pudendal nerve mediated contraction of the urethral sphincters prevents urine leakage under normal conditions. Sensory information from the bladder triggers the micturition reflex leading to bladder emptying. This phase is characterized by the inhibition of the pudendal nerve and suppression of sympathetic activity [13]. Consequently, the detrusor muscle contracts while the pelvic floor muscles and the urethral sphincters relax. When the bladder afferent pathways (from the peripheral pudendal nerves to the spinal cord, and through mesencephalus) stimulate the cerebral cortex, the detrusor center of this region allows micturition to begin or delay [14]. The micturition starts with the external urethral sphincter relaxation, induced by the cerebral cortex, and the detrusor contraction stimulated by the sacral micturition center (located at the spinal cord level S2 to S4) through the pelvic nerves which release acetylcholine to the muscarinic receptors of the detrusor muscle [13, 14]. During the micturition, the simultaneous relaxation of the external sphincter, when the detrusor contracts, is under the control of the pontine micturition center. Therefore, various areas of the nervous system, located in different sites, exert specific control on the bladder and urethral sphincters. In the case of neurogenic disorders, functional changes of detrusor-sphincter complex, clinical signs, and symptoms related, will depend on the site involved by the disease and to the loss of its specific function, as reported in the following scheme.

Lesions above the brain stem: Specific damage in this area causes the loss of the inhibitory control exerted by the cerebral cortex on bladder reflex. As consequence, the detrusor muscle develops uncontrolled function (detrusor overactivity, onset of detrusor involuntary contractions). This altered function may be responsible for the following symptoms: urinary frequency (increase of number of micturitions), urinary urgency (a sudden, compelling urge to urinate), and UI (loss of urine) [15]. In these patients, bladder sensation can be normal to decreased. The urinary sphincters should be synergistic with the bladder (they normally relax when the bladder contracts). Thus, high bladder pressures should not develop and the risk of kidney damage is low.

Complete suprasacral spinal cord lesions: When the lesion is located at the spinal cord above the sacral level, the bladder and sphincter do not receive the control exerted by the cerebral cortex and from the pontine centers. Therefore, the bladder develops uncontrolled activity but, in addition, the urethral external sphincter paradoxically contracts during detrusor activation (dyssynergia), due to the absence of neural mechanisms which provide synergy located in the pontine micturition center [16]. Patients suffering from disorders involving this area may develop the same symptoms of those affected by lesions above the brain stem, but the consequence of sphincter contraction is the high pressure inside the bladder with the risk of vesicoureteral reflux and obstructive bladder voiding with incomplete bladder emptying (poor micturition flow, intermittent flow, urinary retention). Sensation to bladder filling can be normal to decreased. If the lesion is located above T6, the patient may experience autonomic hyperreflexia, which is represented by a constellation of signs/symptoms in response to noxious or non-noxious stimuli below the injury level, including an increase in blood pressure > 20 mmHg above baseline, and may include one or more of following signs and symptoms: headache, flushing and sweating above lesion level, vasoconstriction below lesion level, or dysrhythmias.

Sacral spinal cord lesions: When the sacral level and its micturition center are involved, the bladder reflex is disrupted and patients experience detrusor areflexia. Depending on the type and extent of neurologic injury, decreased bladder compliance may occur during filling. An open smooth sphincter area may result but the striated sphincter may exhibit varied types of dysfunction, although this area usually maintains a resting sphincter tone and cannot be controlled voluntarily [2]. Sensation to bladder filling can be normal to decreased. Therefore, patients express inability to void the bladder with consequent urinary retention, increased risk of urinary infections, bladder stones formation, and hydronephrosis.

Distal to the spinal cord lesions: When this part of the nervous system is damaged, also in these patients the functional alteration leads to detrusor areflexia. However, the smooth sphincter is likely incompetent, and the striated sphincter may exhibit a fixed residual tone that cannot be relaxed voluntarily. Sensation to bladder filling can be normal to decrease [2].

The loss of physiological nervous control of the bladder function causes symptoms that may have a negative impact on a patient's lifestyle and quality of life. While symptoms of the voiding phase may benefit from pharmacological treatment or, in severe cases, from catheterization which can be self-administered at specific time intervals, storage bladder symptoms can hardly limit social interaction, especially when UI is present.

1.3 Onabotulinum toxin type A for treatment of NGB

Treatment of NGB aims to prevent complications to the lower and upper urinary tract secondary to bladder-sphincter dysfunction and consequently improve symptoms and patients' QoL. Major complications of NGB are represented by urinary infections, urinary stones, vesicoureteral reflux, and renal failure. Clinical assessment is based on symptom evaluation, physical examination, renal and bladder ultrasound, and specific instrumental tools, especially flowmetry and urodynamic investigation combined with urethral sphincter electromyography. These tools allow to establish the type and severity of bladder-sphincter dysfunction and to choose treatment. As above reported, UI is the most bothersome symptom. When clinical and urodynamic assessment allows diagnosing NDO as the cause of low bladder compliance and symptoms such as urinary urgency and UI, treatment has the goal to reduce involuntary bladder contractions thus achieving a stable detrusor function (also inducing bladder areflexia). When treatment is effective, intravesical pressure is reduced, and consequently, the risk of vesicoureteral reflux and UI is lower or completely recovered. Various therapeutic options are today available to cure NDO. Based on International Consultation on Incontinence (ICI) algorithms, a conservative approach, followed by oral administration of drugs is recommended as initial management for UI of neurogenic origin, associated with CIC in case of significant post-void residue [17]. Recommended oral drugs are considered antimuscarinics and beta-3-agonists. However, when these therapeutic options fail, the ICI specialized management algorithm suggests BoTA injections into the detrusor muscle for NDO and into the urethral sphincter for bladder-sphincteric dyssynergia. Botulinum toxin causes muscle relaxation (flaccid paralysis) because it binds, at presynaptic level, high-affinity sites on the cholinergic nerve terminals decreasing the release of acetylcholine. After its administration, presynaptic vesicles cannot release the acetylcholine in the synaptic space and consequently, the muscle does not contract. Currently, four different formulations of botulinum toxin, three BTX-A and one botulinum toxin B (BTX-B) are commercially available in Europe and USA: onabotulinumtoxin A / BoTA (Botox, A, Allergan Inc., Irvine, USA) abobotulinumtoxinA (Dysport, Ipsen Limited, Paris, France), incobotulinumtoxinA (Xeomin, Merz Pharmaceutical Raleigh, USA) and rimabotulinumtoxinB (Neurobloc/Myobloc, Solstice Neuroscience Inc., San Francisco, USA). However, adequate clinical data are available only on both BoTA and abobotulinumtoxin B as a treatment option for NDO [18]. Therefore, Literature and international guidelines provide recommendations only for these two formulations to treat NGB, specifically NDO [19]. These two formulations are not interchangeable and of course have different dosing [20], as it's generally accepted that a dosage of 200–300 U of BoTA is comparable with 500–750 U of abobotulinumtoxin [21]. However, further studies have clearly shown that there are no better outcomes comparing both 750 U of abototulinumtoxin and 300 U of BoTA to 500 U abobotulinumtoxin or 200 U BoTA respectively [21, 22]. Although comparative studies are rare and no studies are available comparing different BTX types in the field of urology, in one small non-randomized cohort study on 26 patients, replacement with abobotulinum after the failure of the first injection with BoTA has been proven to be effective [23]. For this reason, a conversion factor between BoTA and abobotulinumtoxin of 1:2.5 has been suggested by Grosse et al., even if this assumption was not scientifically proven and it's believed that a variable conversion rate of the two toxins between 1:2 and 1:3 is applicable [24]. Although both these products are commonly used in real clinical practice, the only FDA-approved dose and formulation for urological application is 200 U of BoTA. Intradetrusor administration of BoTA is performed under local or general anesthesia, using a rigid or flexible cystoscope. A special needle (maximum depth 4 mm) allows to perform the administration of the toxin, usually subdividing the total dose (200 U) in twenty different sites of the bladder, avoiding the bladder dome (to prevent the risk of extra-vesical diffusion) and the trigone (to prevent the risk of vesicoureteral reflux). Large evidence of the safety and efficacy of BoTA on NDO is provided by the literature. A recent review performed by L.F. Cooley and S. Kielb reported long-term data from clinical trials and real-life studies, which show that patients with NDO and detrusor sphincter dyssynergia benefit significantly from intradetrusor BoTA injection with regard to the following parameters: improved voided volume, improved bladder pressure and urodynamic outcomes, reduced incidence of urinary tract infections, and improved QoL [1]. The most important studies providing high-quality data are those from Cruz (2011) and Ginsberg (2012) [25, 26]. In these placebo-controlled protocols, the population in both trials was represented by patients suffering from MS and SCI, with urodynamic evidence of NDO, and submitted to BoTA intradetrusor injections (200 or 300 U). The positive results shown by these studies were confirmed by Kennelly et al. who conducted a 3-year prospective study in 396 patients with SCI and MS to assess long-term efficacy of BoTA injections for NDO [27]. Data on detrusor-sphincter dyssynergia are not so strong and consistent as seen with NDO because outcomes come from low-powered studies. However, a recent meta-analysis of BoTA use in SCI patients suffering from detrusor-sphincter dyssynergia did point to the potential efficacy of this approach with an average decrease from 251.8 mL to 153 mL of post-void residue up to 6 months post-BoTA injection as well as a reduction in sequelae of urinary tract infections and need for CIC in some studies [28]. BoTA is injected on the external urethral sphincter usually through the transperineal way under transrectal ultrasound guidance, and in some cases using electromyography support.

1.4 Quality of life in patients with NGB and improvement after BoTA injection

Restoring the adequate quality of life should be considered a goal of treatment as significant as the recovery from the disease. QoL is what patients experience on a day-to-day basis and is one of the key considerations when they are involved in choices about their medical care [29]. Quality of life was defined in 1947 by the World Health Organization as a "state of complete physical, mental and social well-being, and not merely the absence of disease and infirmity" [30]. In neuro-urology, treatments aim to correct bladder-sphincteric dysfunction to both prevent severe complications to the urinary tract and improve QoL. However, in daily clinical practice (even more so in the past) QoL has been wrongly interpreted as a secondary consideration with respect to the treatment of bladder dysfunction with serious urologic complications and the preservation of renal function. In recent years, especially since 2010, various studies have been published reporting objective data collected by specific patient-reported outcome measures. These studies clearly show how QoL changes during neurologic diseases when bladder symptoms develop [29]. It is important to assess and understand how a person's life is affected by bladder changes that can accompany the neurologic disease, especially UI because this evaluation is directed to target the therapy most effectively. Despite it being one of the fundamental aspects of neuro-urology, there has been not much research on QoL differences across bladder management choices; this fact may represent a limit for the assessment of specific therapeutic algorithms which optimize the relationship between clinical success and QoL improvement for the patient [31]. In fact, while literature provides considerable data to support the improvement of functional outcomes of treatments for NGB (i.e. bladder augmentation), there is a much smaller body of studies supporting objective improvements in QoL [31]. A large amount of research has been focused on the identification of specific QoL tools for neurogenic bladder function after SCI. Best et al. conducted a systematic review of literature published from 1950 to 2015 on this topic and found 42 studies including 24 QoL outcome tools (ten objective, fourteen subjective) [32]. This important review concluded the existing outcome measures representative of the three major domains of QoL as "Achievements", "Utility", and "Subjective well-being". The Authors explain that both objective and subjective measures are important in SCI, concluding that the only validated condition-specific outcomes that show sensitivity to NGB are the QLI-SCI (Quality of Life in Spinal Cord Injury) and Qualiveen, while the SF-36 questionnaire provides a valid assessment of objective QoL in SCI. There are some specific studies regarding QoL of neuro-urologic patient sub-categories. In these papers, the authors point out the importance of using dedicated and validated tools to investigate objective and subjective aspects, including psychometric evaluations. Catherine Browne et al. in 2015 provided a report on QoL for people with MS [33]. Bladder dysfunction has been described in approximately 75% of people with MS [34] and therefore, in this study, participants were recruited from one branch of the Multiple Sclerosis Society of Ireland using purposive sampling techniques [35]. Patients from this cohort (19 subjects, 11 females and 8 males) suffered from at least one urinary symptom: involuntary leakage of urine, voiding frequency (>8), nocturia, voiding dysfunction such as hesitancy, straining, poor stream. Outcomes from this investigation showed that bladder dysfunction creates a sense of disruption and loss for people with MS, interfering with daily activities. One of the most important factors conditioning lifestyle was the unpredictability of bladder symptoms, especially urinary urgency which drives UI. In patients suffering from MS, as other people are affected by neurological impairment, bladder dysfunction is magnified due to other co-occurring symptoms; specifically, physical limitations heighten LUTS in terms of mobility issues creating problems in managing urinary frequency and urgency, often leading to UI. In fact, this study showed that also in MS patients, UI represents the most bothersome symptom associated with emotional consequences also because fear of leaking urine in public may be greater than the distress caused by the leakage of urine itself [36]. Urinary incontinence is prevalent also in SCI individuals. In fact, more than 80% of these subjects experience NGB resulting from neurological impairments that determine NDO +/− sphincter dyssynergia or detrusor areflexia [37]. Spinal cord injury patients are at high risk of complications due to the development of vesicoureteral reflux (in case of detrusor-sphincter dyssynergia and consequent high intra-vesical pressures) or high bladder residual volumes (due to areflexia). These conditions may be responsible for urinary infections, hydronephrosis, finally leading to chronic renal failure, thus requiring strong medical intervention which also can contribute to lifestyle changes. For this reason, NGB remains the most important issue in QoL of patients with SCI, apart from physical movement, and it requires an aggressive attitude towards urinary management in order to improve QoL. As previously reported generically for NGB and SCI patients, the Qualiveen questionnaire has proven to adequately assess disease aspects of limitations, constraints, fears, and feelings [38]. Lundqvist et al. found that UI reduced self-reported QoL among individuals with SCI [39]. The same findings were reported by Westgren and Levi, who described lower QoL in SCI subjects with bladder problems with respect to controls [40]. When the bladder is properly managed, LUTS improve, renal function is preserved, and the person with SCI can enjoy a much healthier life [41]. This outcome has been seen especially when significant improvement in urinary continence has been restored and reported as better body image perception and independence [28]. A significant rate of SCI patients practices self-catheterization to void the bladder due to areflexia, which can be a direct consequence of the spinal lesion or the effect of treatment (antimuscarinics or bladder injections with BoTA). Long-term clean intermittent catheterization (CIC) was first promoted in the 1970s by Lapides et al. [42] and became the standard procedure for managing the NGB of SCI patients [43]. Studies on QoL of patients using CIC show that it has many beneficial effects, which include reduced morbidity and mortality, improved body image, and guaranteed improved self-esteem [44]. These outcomes are even more positive when the use of CIC is associated with a complete recovery of UI (when patients are totally dry), as provided in a large rate of

SCI patients by BoTA administration [45]. These outcomes in SCI submitted to CIC are supported by other studies. Fuminicelli et al. showed that in the Brazilian and Portuguese populations of SCI patients, QoL scores improved in those using CIC because of better independence, self-confidence, social relationship, and access to work activities [46]. The same authors performed a review on the topic including 13 high-quality studies (from the initial 2945 examined) examining QoL assessment in patients with NGB secondary to different disorders (SCI, MS, Parkinson's disease, cerebrovascular accident, brain tumors, infection by HTLV-I, neuroschistosomiasis). The report concluded that CIC offers considerable changes in the NGB patients' living activities, modifying social routines, professional activities, and sexuality, among other areas. However, also in this review, an important concept is the significant role of the recovery from UI to achieve a "dry-status" of the patient using CIC, therefore enhancing the role of proper treatments for the neurogenic UI, as BoTA injections [46]. Intravesical BoTA is a safe and generally well-tolerated procedure because it can be performed under local, regional, or general anesthesia and it requires short operative time. This aspect is important for ensuring good patient acceptance of this treatment which can generally be repeated over time. Quality of life outcomes in patients injected with BoTA has been considered in the most important trials since 2011. Francisco Cruz et al. reported a significant reduction of UI episodes and improvement of urodynamic parameters in 275 patients with NDO after BoTA injection during a multicenter, double-blind, randomized, placebo-controlled trial [47]. This cohort was interviewed during the study by means of Incontinence Quality of Life (I-QOL) questionnaire. Final results showed that among patients submitted to BoTA, 38% and 39,6% respectively submitted to 200 and 300 U were fully dry versus only 7,6 of those in the placebo arm. This clinical finding was associated with a significant improvement of I-QOL total summary scores at 6 weeks from treatment in patients injected, despite the incidence of adverse events (urinary tract infections and urinary retention). Specifically, the total I-QOL score improved from 24.4 to 25.1 in subjects injected with 200 BoTA U, and from 24.3 to 25.9 in those injected with 300 U, while subjects in the placebo group presented a worse score (decreased from 11.7 to 8.6). In the long-term, multi-center, double-blind, randomized, placebo-controlled trial conducted by Ginsberg et al., patients were followed up to 52 weeks. In this protocol, patients were evaluated every 6 weeks after the first 3 months from the first injection of 200 or 300 U, until re-treatment [48]. As shown in the previous trial, also in this study good QoL outcomes were associated with urodynamic improvement. Cystometric measures improved both in SCI and MS subjects, without significant difference between the active dose groups (200 or 300 U). Each BoTA dose significantly improved the I-QOL summary score at week 6 compared with placebo and this result was maintained through week 12 in the overall population with increases from baseline of 9 in the placebo arm, 31 and 33 points in the BoTA 200 and 300 U groups ($p < 0.001$). The most important side effects reported in this trial were UTI and the need of CIC. However, the change from baseline I-QOL score was analyzed in patients who did not perform CIC at baseline to determine whether subsequent initiation of CIC influenced QoL. In these patients, the I-QOL improvement was similar whether they did or did not begin CIC after treatment. The authors commented in the discussion that the onset of BoTA action was rapid and sustained with a duration of effect time approximately of 9 months, probably influencing the positive effect also on I-QOL scores. Particularly, the positive effect of BoTA on recovery of continence is crucial for these patients. Satisfaction with life has been shown to be significantly lower among neurogenic patients with continence problems and the use of BoTA is often required by these subjects because they frequently discontinue antimuscarinics due to failure of response, side effects, or unmet treatment expectations [47]. Sussman et al. in 2012 randomized patients to intradetrusor placebo or BoTA 200 and 300 U. The Patient Report Outcomes included I-QOL to assess Health Related Quality of Life, the 16-item modified Overactive Bladder-Patient Satisfaction with Treatment Questionnaire (OAB-PSTQ) to assess treatment satisfaction, and Patient Global Assessment to assess treatment goal achievement [48]. Patients of the 200 and 300 U groups had improvement of I-QOL scores significantly greater compared with placebo. Improvements were reported in the avoidance/limiting behavior, psychosocial impact, and social embarrassment domain ($p < 0.01$). As observed in other trials, there were no clinically relevant differences between BoTA dose groups for the effects on total I-QOL domain scores. A significant association was then found between UI frequency and improvement of HRQoL. Also, OAB-PSTQ showed that patients treated with BoTA were somewhat or very satisfied with treatment compared to unsatisfied patients of the placebo group. These results were observed in patients regardless of the need for CIC after BoTA injection. Consequently, the final PGA measures indicated that the majority of patients treated with BoTA improved in symptoms, QoL, activity limitations, and overall emotions related to their bladder problems at weeks 6 and 12, and this improvement was maintained or continued to improve during the follow-up. Good outcomes can be achieved in NGB when proper assessment is performed (based on standardized symptoms and QoL questionnaires, clinical assessment, instrumental evaluation by urodynamics). In addition, it is important to follow international guidelines to establish treatment starting as soon as possible to prevent complications and worsening of the bladder-sphincteric dysfunction. In fact, a positive effect of continuous care intervention on the QoL of NGB patients with SCI was observed [49]. The goal of continuous care intervention is to maintain continence, prevent urological complications, and preserve upper urinary tract function to make bladder management compatible with the person's lifestyle and environment [50]. In patients submitted to BoTA injections, continuous care intervention may be an important method of re-assessment over time to suggest the need of instrumental re-evaluation and re-treatment. A significant part of continuous care is represented by training for CIC and adequate suggestions which prevent infections. It was observed that continuous care interventions contribute to improve the QoL of patients after 3 months. This is the consequence of better bladder management with regard to urinary system complications.

2. Conclusions

Patients suffering from neurogenic disorders adapt to physical and social limitations. The effect of physical disability of illness cannot be understood if QoL aspects of importance for the individual are not taken into consideration. When urinary symptoms and bladder-sphincteric dysfunction develop, these patients have a significant worsening of their QoL. This report clearly shows that UI is the most bothersome symptom to manage in this population and that it is associated with a hard negative impact on QoL. Recovery of continence allowed by intradetrusor injection of BoTA provides a great QoL improvement which is parallel to the bladder-sphincteric functional modification. This result is supported by various multicentric, randomized, placebo-controlled trials with a specific evaluation of QoL by means of standardized and highly recommended patients' report outcomes. These studies guarantee that good outcomes are maintained over time despite the need to start CIC. For this reason, BoTA injections are strongly recommended by International Guidelines in these patients. In the future, the authorized use of different types of botulinum toxin for urological use is expected, extending the indication to the pediatric population. Furthermore, large multi-center studies are warranted to design specific protocols which should guide clinicians in managing NGB patients who need re-treatments (in terms of time intervals and BoTA dosing), and which can support the management of subjects who are refractory to BoTA, replacing it by different toxins. However, considering that QoL can be influenced by diverse factors and not only by treatments, it is important to remember that family support, adjustment and coping, productivity, self-esteem, financial stability, education, and physical and social environments must also be assessed and considered in NGB individuals.

Author details

Giovanni Palleschi[1]* and Antonio Cardi[2]

1 Medica San Carlo Nephrology, Urology and Hemodialysis Center, Frascati (Rome), Italy

2 San Giovanni Addolorata Hospital, Rome, Italy

*Address all correspondence to: giovanni.palleschi@uniroma1.it

References

[1] Cooley LF, Kielb S. A review of botulinum toxin A for the treatment of neurogenic bladder. PMR Feb 2019;11(2):192-200

[2] Dorsher PT, McIntosh PM. Neurogenic bladder. Advances in Urology. 2012;**2012**:816274

[3] Lansang RS, Krouskop AC. In: Massagli TL et al., editors. Bladder management. London: Nature Publishing Group, eMedicine; 2004

[4] Verhoef M, Lurvink M, Barf HA, et al. High prevalence of incontinence among young adults with spina bifida: Description, prediction and problem perception. Spinal Cord. 2005;**43**:331-340

[5] Ginsberg D. The epidemiology and pathophysiology of neurogenic bladder. The American Journal of Managed Care. 2013;**19**:S191-S196

[6] Shamliyan TA, Wyman JF, Ping R, Wilt TJ, Kane RL. Male urinary incontinence: Prevalence, risk factors, and preventive interventions. Revista de Urología. 2009;**11**(3):145-165

[7] Ueda T, Tamaki M, Kageyama S, Yoshimura N, Yoshida O. Urinary incontinence among community-dwelling people aged 40 years or older in Japan: Prevalence, risk factors, knowledge and self-perception. International Journal of Urology. 2000;7:95

[8] Finkelstein MM. Medical conditions, medications and urinary incontinence. Analysis of a population-based survey. Canadian Family Physician. 2001;**48**:96-101

[9] Østbye T, Borrie MJ, Hunskaar S. The prevalence of urinary incontinence in elderly Canadians and its association with dementia, ambulatory function, and institutionalization. Norsk Epidemiologi. 2009;**8**(2):177-182

[10] Huang AJ, Brown JS, Thom DH, Fink HA, Yaffe K, Study of Osteoporotic Fractures Research Group. Urinary incontinence in older community-dwelling women: The role of cognitive and physical function decline. Obstetrics and Gynecology. 2007;**109**(4):909-916

[11] Hinman F Jr. Nonneurogenic neurogenic bladder (the Hinman syndrome) 15 years later. The Journal of Urology. 1986;**136**(4):769-777

[12] Ginsberg D. The epidemiology and pathophysiology of neurogenic bladder. The American Journal of Managed Care. 2013;**19**(10 Suppl):s191-s196

[13] Fowler CJ, Griffiths D, de Groat WC. The neural control of micturition. Nature Reviews. Neuroscience. 2008;**9**:453-466

[14] Al-Shukri SA. Neurogenic bladder-assessment, investigation, and treatment. European Urological Review. 2012;7:55-60

[15] Abrams P, Cardozo L, Fall M, Griffiths D, Rosier P, Ulmsten U, et al. The standardisation of terminology in lower urinary tract function: Report from the standardisation sub-committee of the international continence society. Urology. 2003;**61**(1):37-49

[16] Wein AJ, Dmochowski RR. Neuromuscular dysfunction of the lower urinary tract. In: Wein AJ, Kavoussi LR, Novick AC, et al., editors. Campbell-Walsh Urology. 10th ed. Philadelphia, PA: Elsevier Saunders; 2010. pp. 1909-1946

[17] Abrams P, Cardozo L, Wagg A, Wein A. Incontinence. 6th ed. ICI-ICS. Bristol UK: International Continence Society; 2017. ISBN: 978-0956960733

[18] Weckx F, Tutolo M, De Ridder D, Van der Aa F. The role of botulinum toxin A in treating neurogenic bladder. Translational andrology and urology. 2016;**5**(1):63-71

[19] Chancellor MB, Elovic E, Esquenazi A et al. Evidence-based review and assessment of botulinum neurotoxin for the treatment of urologic conditions. Toxicon. 1 Jun 2013;17:115-128

[20] Mangera A, Andersson KE, Apostolidis A, et al. Contemporary management of lower urinary tract disease with botulinum toxin A: A systematic review of botox (OnabotulinumtoxinA) and disport (abobotulinumtoxinA). European Urology. 2011;**60**:784-795

[21] Grise P, Ruffion A, Denys P, et al. Efficacy and tolerability of botulinum toxin type A in patients with neurogenic detrusor overactivity and without concomitant anticholinergic therapy: Comparison of two doses. European Urology. 2010;**58**:759-766

[22] Mohee A, Khan A, Harris N, et al. Long term outcome of the use of intravesical botulinum toxin fot the treatment of overactive bladder (OAB). BJU International. 2013;**111**:106-113

[23] Peyronnet B, Roumiguié M, Caste-Lacanal E, et al. Preliminary results of botulinum toxin A switch after first detrusor injection failure as a treatment of neurogenic detrusor overactivity. Neurourology and Urodynamics. 2016 Feb;**35**(2):267-270

[24] Grosse J, Kramer G, Storher M. Success of repeat detrusor injections of botulinum toxin in patients with severe neurogenic detrusor overactivity and incontinence. European Urology. 2005;**47**:653-659

[25] Cruz F, Hershorn S, Aliotta P, et al. Efficacy and safety of onabotulinumtoxinA in patients with urinary incontinence due to neurogenic detrusor overactivity: A randomized, double blind, placebo controlled trial. European Urology. 2011;**60**:742-750

[26] Ginsberg D, Gousse A, Keppenne V, et al. Phase 3 efficacy and tolerability study of onabotulinumtoxinA for urinary incontinence from neurogenic detrusor overactivity. The Journal of Urology. 2012;**187**:2131-2139

[27] Kennelly M, Dmochowski R, Ethans K, et al. Long term efficacy and safety of onabotulinumtoxinA in patients with urinary incontinence due to neurogenic detrusor overactivity: An interim analysis. Urology. 2013;**81**:491-497

[28] Benner JS, Nichil MB, Rovner ES, et al. Patient reported reasons for discontinuing overactive bladder medication. BJU International. 2010;**105**:1276-1282

[29] Welk B. Quality of life in neurourology patients. European Urology Focus. 2020;**6**(3):531-533

[30] Post m. Definitions of quality of life: What has happened and how to move on. Topics in Spinal Cord Injury Rehabilitation. 2014;**20**:167-180

[31] Myers JB, Lenherr SM, Stoffel JT, et al. Patient reported bladder related symptoms and quality of life after spinal cord injury with different bladder management strategies. The Journal of Urology. 2019;**202**:574-584

[32] Best KL, Ethans K, Craven BC, Noreau L, Hitzig SL. Identifying and classifying quality of life tools for neurogenic bladder function after spinal cord injury: A systematic review. The Journal of Spinal Cord Medicine. 2017;**40**(5):505-529

[33] Browne C, Salmon N, Kehoe M. Bladder dysfunction and quality of life for people with multiple sclerosis. Disability and Rehabilitation. 2015;**37**(25):2350-2358

[34] Norvedt M, Frugaard J, Mohn J, et al. Prevalence of bladder, bowel and sexual problems among multiple sclerosis patients two to five years after diagnosis. Multiple Sclerosis. 2007;**13**:106-112

[35] Patton MQ. Qualitative Evaluation and Research Methods. 2nd ed. Newbury Park (CA): Sage Publications; 2013

[36] Kayes NM, McPherson KM, Schluter P, et al. Exploring the facilitators and barriers to engagement in physical activity for people with multiple sclerosis. Disability and Rehabilitation. 2011;**33**:1043-1053

[37] Tawel WA, Seyam R. Neurogenic bladder in spinal cord injury patients. Research and Reports in Urology. 2015;7:85-99

[38] Hill VB, Davies WE. A swing to intermittent clean self-catheterization as a preferred mode of management of the neuropathic bladder for the dexterous spinal cord patient. Paraplegia. 1988;**26**:405-412

[39] Lundqvist C, Siosteen A, Blomstrand C, Lind B, Sullivan M. Spinal cord injuries. Clinical, functional and emotional status. Spine. 1991;**16**:78-83

[40] Westgren N, Levi R. Quality of life and traumatic spinal cord injury. Archives of Physical Medicine and Rehabilitation. 1998;**79**:1433-1439

[41] Ku JH. The management of neurogenic bladder and quality of life in spinal cord injury. BJU International. 2006;**98**:739-745

[42] Lapides J, Diokno AC, Silber SJ, Lowe BS. Clean, intermittent catheterization in the treatment of urinary disease. The Journal of Urology. 1972;**107**:458-461

[43] Kuhn W, Rist M, Zaech GA. Intermittent urethral self catheterization: Long term results (bacteriological evolution, continence, acceptance, complications). Paraplegia. 1991;**29**:222-232

[44] Weld KG, Dmochowski RR. Effect of bladder management on urological complications in spinal cord injured patients. The Journal of Urology. 2000;**163**:768-772

[45] Schurch B, de Seze M, Denys P, et al. Botulinum toxin type A is safe and effective treatment for neurogenic urinary incontinence: Results of a single treatment, randomized, placebo, controlled 6 month study. The Journal of Urology. 2005;**174**:196-200

[46] Fuminicelli L, Mazzo A, Martins JCA, Dias Henriques FM, Orlandin L. Quality of life of patients using intermittent catheterization. Revista Latino-Americana de Enfermagem. 2017;**25**:e2906

[47] Benner JS, Nichil MB, Rovner ES, et al. Patient reported reasons for discontinuing overactive bladder medication. BJU International. 2010;**105**:1276-1282

[48] Sussman D, Patel V, Del Popolo G, Lam W, Globe D, Pommerville P.

Treatment satisfaction and improvement in health-related quality of life with OnabotulinumtoxinA in patients with urinary incontinence due to neurogenic detrusor overactivity. Neurourology and Urodynamics. 2013;**32**(3):242-249

[49] Tang F, Cheng Z, Wen X, Guan J. Effect of continuous care intervention on the quality of life in patients with neurogenic bladder dysfunction. Journal of International Medical Research. 2017;**47**(5):2011-2017

[50] Yang CC. Bladder management in multiple sclerosis. Physical Medicine and Rehabilitation Clinics of North America. 2013;**24**:673-686

Chapter 7

Clinical Relevance of Neutralizing Antibodies in Botulinum Neurotoxin Type A

Harald Hefter and Sara Samadzadeh

Abstract

The precise definition of prevalence of neutralizing antibodies (NABs) affords cross-sectional testing of a cohort. But in most studies, only selected patients are tested. This leads to gross underestimation of NAB-prevalence, and the opinion that induction of NABs is a rare phenomenon in botulinum neurotoxin (BoNT)/A-therapy. However, recent cross-sectional studies report annual incidences between 1 and 2% in patients being treated with a complex protein (CP)-containing preparation. This implies that NAB-prevalence above 10% has to be expected in patients being treated for more than 10 years. High dose per session and long duration of treatment are relevant risk factors for induction of NABs. In patients exclusively treated with the CP-free incobotulinumtoxin A (incoBoNT/A) preparation Xeomin® no NAB-induction has been reported so far. In patients with NABs switching to incoBoNT/A may lead to a decline of NAB-titers. In patients with NABs under treatment with a CP-containing BoNT/A-preparation it may take years of treatment until a second treatment failure (STF) becomes clinical manifest. In a cohort of 59 patients with partial STF patients' reports on the reduction of BoNT-activity predicted the presence of NABs better than treatment related data produced by the treating physicians.

Keywords: botulinum neurotoxin, neutralizing antibodies, immunogenicity, long-term treatment, treatment failure, secondary nonresponse

1. Introduction

Botulinum neurotoxins (BoNTs) are bacterial products and large molecules which are usually embedded into even larger complex proteins (CPs) [1–4]. In clinical practice, BoNTs have to be applied by injection. This causes activation of local dendritic cells, elicits hit-shock proteins, and leads to local inflammation. Therefore, the induction of antibodies (ABs) can hardly be avoided [5]. The analysis of lymphocytes in BoNT/A or BoNT/B long-term treated patients with movement disorders indicates that in most of these patients, T-cells have responded to the BoNT application [6].

Induced antibodies target epitopes of CPs and BoNT. Some of the ABs do not influence the biological activity of BoNTs, and others reduce or neutralize BoNT action [7]. Neutralizing antibodies (NABs) in immune-resistant patients target epitopes of the heavy [8] or the light chain [9] of the BoNT molecule.

In clinical practice, the relevant question arises whether a partial or complete secondary treatment failure (pSTF or cSTF) results from NAB induction.

2. The problem of precise determination of the prevalence of ABs and/or STF

For the determination of the presence of NABs, clinical as well as laboratory tests can be used (for a recent discussion, see [10]). Regardless of which test is addressed, little is known about the test-retest liability or variability of repeated measurements of a single serum or serial measurements within a single patient. Nevertheless, the precise determination of the prevalence of ABs heavily depends on the quality (sensitivity and specificity) of NAB testing.

Furthermore, by definition of prevalence, it is necessary to test all members of a cohort if the prevalence of a certain feature in a special cohort has to be determined. Thus, precise determination of the prevalence of neutralizing antibodies in a cohort of BoNT-treated patients implies that a cross-sectional NAB study has to be performed in this cohort.

In the majority of studies reporting on the prevalence of NABs, no cross-sectional testing has been performed. Instead, antibody rates or antibody frequencies (= number of NAB-positive patients/number of patients in the cohort) are determined resulting from NAB testing of selected patients. This procedure crossly underestimates the presence of NABs (comp. [11]).

In their meta-analysis of NABs in BoNT therapy, Fabbri et al. report on an overall NAB frequency of 3.5% in clinically responding patients. In responding patients with spasticity, only 0.7% were reported to be NAB-positive, in patients with dystonia 6.5%. These data are at clear variance with cross-sectional studies reporting 31 positive patients among 212 responding patients with cervical dystonia who were tested by means of the mouse hemidiaphragma assay (MHDA) (= 14.6%; [12]) and 14.3% MHDA-positive patients in long-term treated patients with spasticity [13].

An even more challenging problem is the determination of the temporal development of NAB prevalence. Antibody rates or antibody frequencies do not give any information on the temporal development of NABs, since this ratio between NAB-positive patients and all members of the cohort does not take into account the duration of BoNT treatment.

To analyze the temporal development of NAB formation, the Kaplan-Meier survival analysis has to be performed, calculating the prevalence of NAB-positive patients among those patients with a given duration of treatment. This approach demonstrates that up to 50% of the patients will become NAB-positive when treatment durations exceed 25 years or higher doses are used. In **Figure 1**, the probability to remain AB-negative is plotted against the duration of treatment in 595 patients. The prevalence of NABs was 82 out of 594 patients (= 13.8%) in the entire cohort, 7 out of 186 patients (= 3.8%) being treated with doses <500 uDU, 49 out of 312 patients (= 15.7%) being treated with doses between 500 and 1000 uDu, and 26 out of 96 patients (= 27%) when doses larger than 1000 uDU were used. To compare doses of different BoNT/A preparations and to determine unified dose units (uDU), aboBoNT/A doses were left unchanged and ona- and incoBoNT/A doses were multiplied by 3. These conversion ratios have been used by Fabbri et al. 2016 [11] and have been discussed by Contarino et al. [14]. The Kaplan-Meier analysis clearly reveals

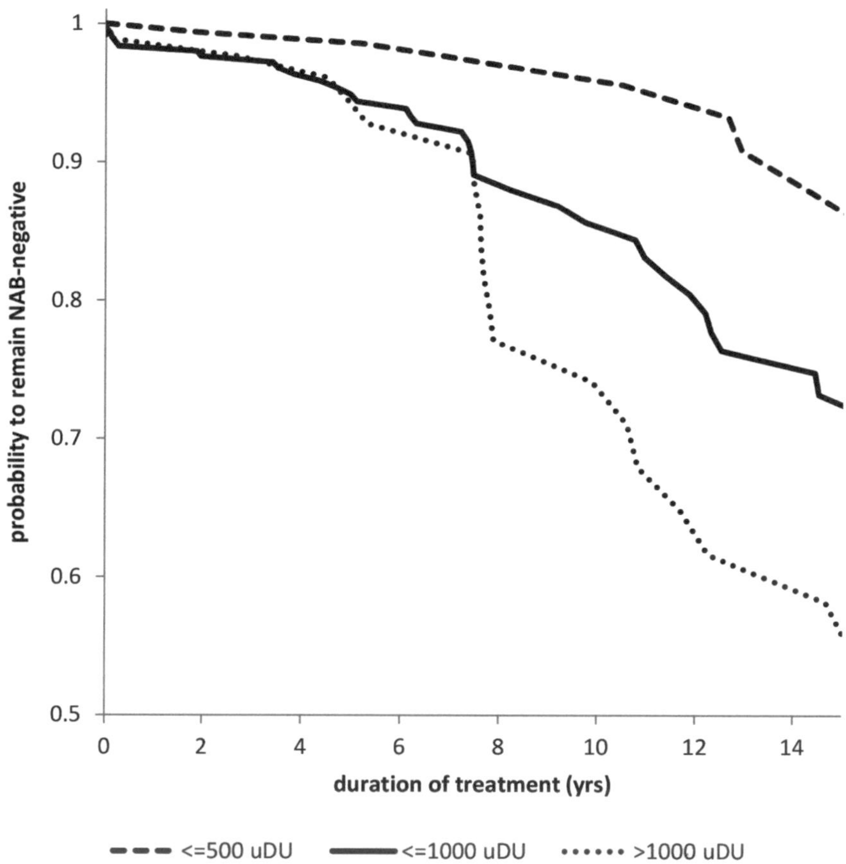

Figure 1.
Kaplan-Meier survival curve to remain AB-negative in patients with cervical dystonia being treated with BoNT/A. With increasing duration of treatment beyond 10 years, the probability to remain AB-negative declines rapidly down to values around 50%.

that there is a nonlinear decline of the probability to remain MHDA-negative with the duration of treatment, especially in the higher dose groups.

In summary, induction of NABs occurs frequently and inevitably progresses with the duration of treatment as long as complex protein-containing BoNT/A preparations are used. The induction of NABs is probably significantly lower in complex protein-free BoNT/A preparations [5, 15] (see Section 3 below).

The determination of incidence and prevalence of secondary treatment failure (STF) is even more complex. So far, there is no clear definition of STF. If no response to a BoNT injection can be detected neither by the patient nor the treating physician, the diagnosis of a complete treatment failure (cSTF) is comparably easy [16]. However, this is the end stage of a longer process, starting with an increasing reduction of the duration of action of a BoNT injection before the 4-week peak effect becomes reduced. This has been described in 2004 by Dressler in detail [1].

When patients are reinjected every 3 months, the reduction of the duration of action cannot be measured directly but results in a systematic worsening of disease severity (for details see [5, 17]). We, therefore, have proposed a formal definition of pSTF in patients with CD. If a patient has responded with more than three points on the TSUI scale [18] and then develops a systematic worsening over three injection cycles of more than two TSUI score points and reports a reduction of the effect of BoNT injection to head position, tremor, or pain, a pSTF has to be suspected.

Our experience is that such a formal definition of pSTF is helpful to detect patients with pSTF. Furthermore, in a cohort of 32 patients with pSTF according to this definition, 25 patients (= 78%) had a positive MHDA test [4]. This is definitely more than in a large cohort of patients with suspected pSTF without formal definition were only about 50% were MHDA-positive [19].

This led to the opinion that NAB-associated secondary nonresponse (SnR) is a rare occurrence, and SnR is more frequently due to an insufficient dose, inappropriate muscle selection, or improper injection technique or targeting [10, 20]. But to our opinion, these aspects of BoNT treatment do not suggest the development of pSTF but indicate insufficient and inappropriate BoNT treatment. pSTF can only be suspected when the patient worsens despite of therapy optimization.

In a cross-sectional study on 66 patients with CD who had started their BoNT therapy with abo- or onaBoNT/A, we have analyzed patients′ drawing of the course of disease (course of disease graphs (CoDGs)) after the onset of BoNT therapy over the entire duration of BoNT therapy. Five different response types could be distinguished: the rapid or golden responder (RR) type, the continuous response (CR) type, the poor response (PR) type, and the secondary treatment failure (STF) type I and II.

The RR type is characterized by a rapid response after the onset of BoNT therapy, followed by a further less rapid improvement. The CR type is characterized by a continuous improvement over the entire duration of treatment, The PR type is characterized by an improvement of less than 20%. Patients who drew a PR type CoD graph were primary nonresponders. The STF types I and II are characterized by an initial improvement followed by a secondary worsening. In the STF type II, the second period of improvement followed after the switch of the BoNT preparation.

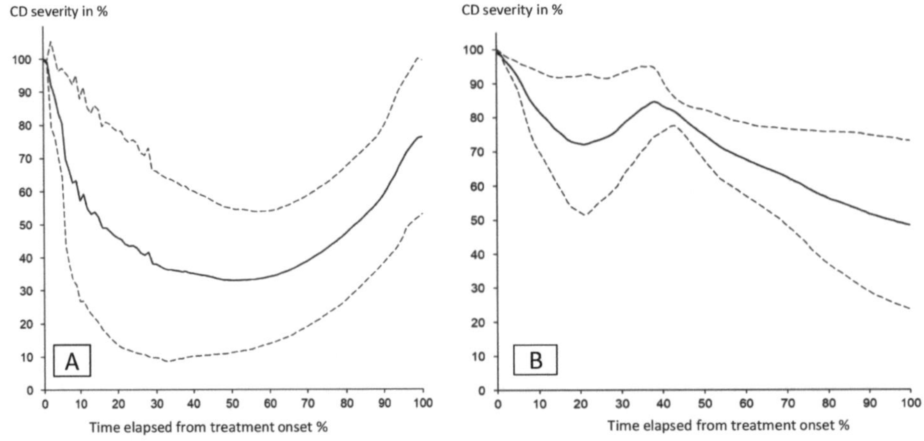

Figure 2.
The severity of cervical dystonia versus percentage of time elapsed from treatment onset. A. Mean course of disease graph (CoD graph; solid line) plus 1 standard deviation range (hatched lines) of 11 CD patients who had drawn a STF type I CoD graph after the onset of BoNT therapy. The severity of CD rapidly decreases initially but then worsens again. B. Mean course of disease graph (CoD graph; solid line) plus 1 standard deviation range (hatched lines) of six CD patients who had drawn a STF type II CoD graph after the onset of BoNT therapy. The severity of CD improved initially, then worsened again, but improved a second time after the switch of the BoNT preparation.

Among the 66 patients, 17 patients (= 25.8%) produced a STF type. In **Figure 2A**, the mean CoD graph (full line) plus/minus 1 standard deviation (hatched lines of the 11 patients with a STF type I CoD graph is presented. In **Figure 2B**, the corresponding mean CoD graph (full line) plus/minus 1 standard deviation of the six STF-type II CoD graphs are presented.

The same task (to draw the course of disease after the onset of BoNT therapy) was analyzed in 34 patients who had exclusively been treated with incoBoNT/A. No patient in the incoBoNT/A monotherapy group produced a STF type I or II CoD graph. This difference in frequency of drawing STF type I or II graphs is significant ($p < .05$).

In summary, partial secondary treatment failure (pSTF) occurs more frequently than positive NAB tests suggest. This indicates that NAB tests are less sensitive to detect pSTF than careful clinical investigation and patient's assessment of the efficacy of BoNT therapy.

3. Implications of the presence of NABs on long-term outcome

As mentioned, earlier, the presence of NABs does not imply that there is no clinical response at all. We have recently described a small cohort of nine CD patients with positive NAB testing in 2010 who did not want to be switched to another BoNT preparation. They were retested in 2013 and 2017. Their relation between TSUI scores and paralysis times in 2010, 2013, and 2017 are presented in **Figure 3A** (open symbols indicate data from 2010, full symbols indicate data from 2017, and missing symbols indicate data from 2013). Apart from two exceptional cases, little changes in paralysis times and TSUI scores can be observed.

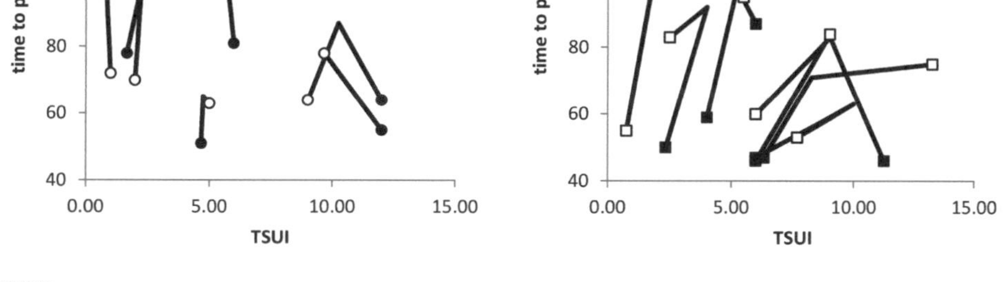

Figure 3.
Time to paralysis versus TSUI. A. Temporal development of the relation between TSUI scores and paralysis times in 2010, 2013, and 2017 in nine CD patients in whom the complex protein-containing preparation had not been switched, although the initial MHDA test was positive (paralysis time > 60 mins). Open circles indicate values of the investigation in 2010, full circles indicate values of the investigation in 2017, and no circles indicate values of the investigation in 2013. B. Temporal development of the relation between TSUI scores and paralysis times in 2010, 2013, and 2017 in nine CD patients in whom the complex protein-containing preparation was switched to incoBoNT/A in 2010. Open squares indicate values of the investigation in 2010, full squares indicate values of the investigation in 2017, and no squares indicate values of the investigation in 2013. Apart from one exceptional case, paralysis times of eight patients decrease between 2010 and 2017. In six out of nine patients, TSUI scores improve, but the improvement of the TSUI score is less pronounced compared to the improvement of the paralysis times.

The data of these nine CD patients in whom BoNT/A preparation had not been switched between 2010 and 2017 were compared to data of nine CD patients in whom the complex protein-containing BoNT/A preparation was switched to the complex protein-free incoBoNT/A preparation in 2010. The relation between TSUI scores and paralysis times of these nine switchers is presented in **Figure 3B**. Apart from one exceptional case, paralysis times improved in the switchers. In parallel, the TSUI scores also improved in six out of nine patients, but the improvement of TSUI scores was less pronounced than the improvement of the paralysis times (see **Figure 3B**).

In summary, so far little is known about the development of NAB titers when patients remain on their BoNT preparation under which they have developed NABs. However, there is increasing evidence that switching from a complex protein-containing preparation (abo- and onaBoNT) to a complex protein-free preparation (incoBoNT/A) may lead to a significant long-lasting improvement of paralysis times [4, 21].

4. No NAB induction in patients under incoBoNT long-term monotherapy

Soon after the treatment of patients with the old "Botox," it became obvious that in a large percentage of patients, NABs were induced [22]. This led to a purification process of the "old" Botox® preparation, a reduction of the protein load of the "new" Botox® by a factor of 5–6 [10, 23], and an improvement of the antigenicity of the Botox® preparation [24]. The incidence of NAB induction reported for the "new" Botox® preparation was around 1%/year [24].

In 2005, incoBoNT was licensed for the treatment of CD [25]. By removal of botulinum neurotoxin complex proteins and elimination of biological inactive fragments of the BoNT molecule, the protein load of this preparation was reduced down to 0.55 ng compared to 4.8 ng of the Botox and 5.3 ng of the Dysport® preparation [3, 26]. But so far, no convincing study has been presented that the lower protein load of the incoBoNT preparation also leads to a significantly lower antigenicity compared to the other two BoNT/A preparations licensed in Europe.

However, it has not been reported that NABs or pSTF were induced in a patient who had exclusively been treated with incoBoNT/A [15]. Cases with NABs have been presented who had been treated only for a few treatment cycles with abo- or onaBoNT and then were switched to incoBoNT/A [1, 15], but NAB induction under incoBoNT/A monotherapy has not been observed.

This is in line with a recent observation that in 34 CD patients who were exclusively been treated with incoBoNT/A, no patient had a positive MHDA test [15] and no patient drew a STF type I or II CoD graph (see Section 2 above).

In summary, the antigenicity of the complex protein-free incoBoNT/A preparation appears to be very low, since no NAB induction has been observed in patients who have exclusively been treated with incoBoNT/A.

5. Prediction of the NAB presence from clinical treatment-related data

It has been demonstrated that long-term treated CD patients with NABs have a significantly worse TSUI score, were treated with significantly higher doses, and have higher pain scores of the CDQ24 than long-term treated CD patients with a negative MHDA [12, 27, 28]. The paralysis times of the MHDA are significantly correlated with the actual doses, the actual TSUI scores, and the CDQ24 pain scores [28].

To analyze which treatment-related parameter may be used to predict the presence of NABs best, NABs were determined in 59 patients with pSTF. Patients 'assessments of the effect of long-term BoNT/A treatment were determined by asking patients to estimate the improvement of CD in percent of the initial severity of CD at the onset of BoNT therapy (IMPQ) and to mark the actual severity of CD in percent of the initial severity on a visual analog scale which yielded a second estimation of improvement (IMPD). The receiver operating characteristics (ROC) curves for the prediction of the presence of NABs by IMPQ and IMPD are presented in **Figure 4A**. The sensitivity and specificity of both parameters were around 0.7–0.8.

The treating physicians scored the actual severity of CD by means of the TSUI score (ATSUI) and determined the improvement since the onset of BoNT therapy by calculating the difference between ATSUI and the initial severity of CD at the onset of BoNT therapy (IMPTSUI) in the same 59 patients. Similar ROC curves were calculated for ATSUI and IMPTSUI (**Figure 4B**). Treating physicians′ scoring predicted the presence of antibodies less well compared to the assessment of the patients. Sensitivity was lower than 0.6, and specificity was also between 0.7 and 0.8.

In summary, patients realize the NAB-induced reduction of the efficacy of BoNT injections quite well, probably better than treating physicians scoring of the treatment effect.

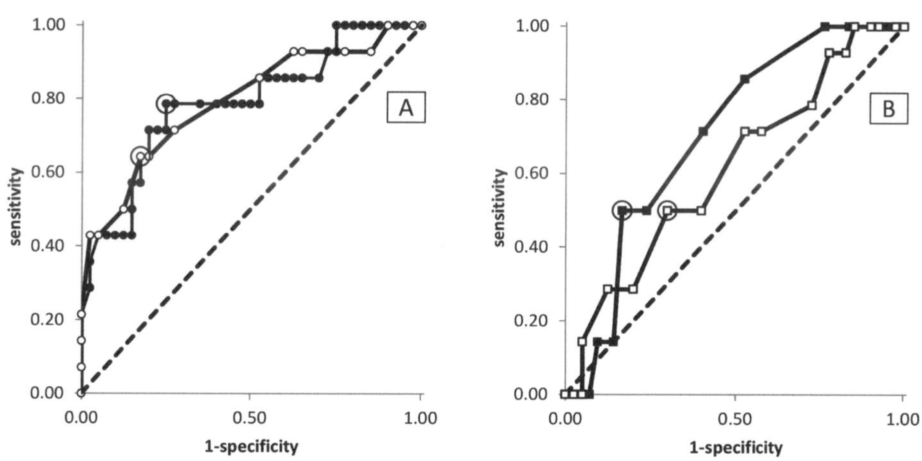

Figure 4.
The ROC curves for prediction of the presence of NABs by IMPQ and IMPD. A. ROC curves analyzing the relation between the presence of NABs and patients′ assessments of the improvement of CD since the onset of BoNT therapy: Solid circles indicate the ROC curve of IMPQ and light circles indicate the ROC curve of IMPD. For both parameters, sensitivity and specificity lie around 0.7–0.8. B. ROC curves analyzing the relation between the presence of NABs and treating physicians' scoring of the actual severity of CD (ATSUI) and the improvement of CD since the onset of BoNT therapy (IMPTSUI): Solid squares indicate the ROC curve of ATSUI and light circles indicate the ROC curve of IMPTSUI. For both parameters, the sensitivity is around 0.5, whereas the specificity lies around 0.7–0.8.

6. Conclusion

Induction of NABs occurs frequently (**Figure 1**), may become manifest after years of successful treatment (**Figure 3**), progresses with the duration of treatment (**Figure 1**), and has clinical implications. In patients with CD, it goes along with higher severity of

CD, leads to more pain, and affords treatment with increasingly higher doses. Patients realize the reduction of efficacy of BoNT/A treatment quite well (**Figure 4**).

Since the induction of NABs has not been observed under monotherapy with incoBoNT/A and switch to incoBoNT/A may lead to clinical improvement in patients with pSTF after ona- and aboBoNT/A incoBoNT/A seems to have a very low antigenicity. We, therefore, recommend using the complex protein-free BoNT/A preparation incoBoNT/A from the very beginning of BoNT/A therapy to reduce the risk of antibody formation as low as possible.

Acknowledgements

We would like to thank the following individuals for their expertise and assistance throughout all aspects of our studies especially Prof. Dr. Philipp Albrecht for making substantial conceptual and design contributions, Dr. Marek Moll, Beyza Ürer, Raphaela Brauns, and Dietmar Rosenthal for contributing to gathering data conducting the data analysis and creating the tables and figures.

Conflict of interest

The authors declare no conflict of interest.

Author details

Harald Hefter and Sara Samadzadeh*
Department of Neurology, University of Düsseldorf, Düsseldorf, Germany

*Address all correspondence to: sara.samadzadeh@yahoo.com

References

[1] Dressler D, Hallett M. Immunological aspects of Botox, Dysport and Myobloc/NeuroBloc. European Journal of Neurology. 2006;**13**(Suppl 1):11-15

[2] Bigalke H. Properties of Pharmaceutical Products of Botulinum Neurotoxins. Botulinum Toxin Therapeutic Clinical Practice & Science. Amsterdam: Elsevier; 2009

[3] Frevert J. Content of botulinum neurotoxin in Botox®/Vistabel®, Dysport®/Azzalure®, and Xeomin®/Bocouture®. Drugs R & D. 2010; **10**(2):67-73

[4] Hefter H, Hartmann CJ, Kahlen U, Samadzadeh S, Rosenthal D, Moll M. Clinical improvement after treatment with IncobotulinumtoxinA (XEOMIN®) in patients with cervical dystonia resistant to botulinum toxin preparations containing complexing proteins. Frontiers in Neurology. 2021;**12**:636590

[5] Samadzadeh S, Ürer B, Brauns R, Rosenthal D, Lee J-I, Albrecht P, et al. Clinical implications of difference in antigenicity of different botulinum neurotoxin type a preparations: Clinical take-home messages from our research pool and literature. Toxins. 2020;**12**(8):499

[6] Oshima M, Deitiker PR, Jankovic J, Duane DD, Aoki KR, Atassi MZ. Human T-cell responses to botulinum neurotoxin. Responses in vitro of lymphocytes from patients with cervical dystonia and/or other movement disorders treated with BoNT/A or BoNT/B. Journal of Neuroimmunology. 2011;**240-241**:121-128

[7] Göschel H, Wohlfarth K, Frevert J, Dengler R, Bigalke H. Botulinum A toxin therapy: Neutralizing and nonneutralizing antibodies--therapeutic consequences. Experimental Neurology. 1997;**147**(1):96-102

[8] Dolimbek BZ, Aoki KR, Steward LE, Jankovic J, Atassi MZ. Mapping of the regions on the heavy chain of botulinum neurotoxin A (BoNT/A) recognized by antibodies of cervical dystonia patients with immunoresistance to BoNT/A. Molecular Immunology. 2007;**44**(5):1029-1041

[9] Atassi MZ, Dolimbek BZ, Jankovic J, Steward LE, Aoki KR. Regions of botulinum neurotoxin A light chain recognized by human anti-toxin antibodies from cervical dystonia patients immunoresistant to toxin treatment. The antigenic structure of the active toxin recognized by human antibodies. Immunobiology. 2011;**216**(7):782-792

[10] Bellows S, Jankovic J. Immunogenicity associated with botulinum toxin treatment. Toxins (Basel). 2019;**11**(9):491

[11] Fabbri M, Leodori G, Fernandes RM, Bhidayasiri R, Marti MJ, Colosimo C, et al. Neutralizing antibody and botulinum toxin therapy: A systematic review and meta-analysis. Neurotoxicity Research. 2016;**29**(1):105-117

[12] Hefter H, Rosenthal D, Moll M. High botulinum toxin-neutralizing antibody prevalence under long-term cervical dystonia treatment. Movement Disorders Clinical Practice. 2016;**3**(5):500-506

[13] Albrecht P, Jansen A, Lee JI, Moll M, Ringelstein M, Rosenthal D, et al. High prevalence of neutralizing antibodies after long-term botulinum neurotoxin therapy. Neurology. 2019;**92**(1):e48-e54

[14] Contarino MF, Van Den Dool J, Balash Y, Bhatia K, Giladi N, Koelman JH,

et al. Clinical practice: Evidence-based recommendations for the treatment of cervical dystonia with botulinum toxin. Frontiers in Neurology. 2017;**8**:35

[15] Hefter H, Brauns R, Ürer B, Rosenthal D, Albrecht P. Effective long-term treatment with incobotulinumtoxin (Xeomin®) without neutralizing antibody induction: A monocentric, cross-sectional study. Journal of Neurology. 2020;**267**(5):1340-1347

[16] Walter U, Mühlenhoff C, Benecke R, Dressler D, Mix E, Alt J, et al. Frequency and risk factors of antibody-induced secondary failure of botulinum neurotoxin therapy. Neurology. 2020;**94**(20):e2109-e2e20

[17] Hefter H, Spiess C, Rosenthal D. Very early reduction in efficacy of botulinum toxin therapy for cervical dystonia in patients with subsequent secondary treatment failure: A retrospective analysis. Journal of Neural Transmission (Vienna). 2014;**121**(5):513-519

[18] TSUI JKC, Stoessl A, Eisen A, Calne S, Calne DB. Double-blind study of botulinumtoxin in spasmodic torticollis. The Lancet. 1986;**328**:245-247

[19] Lange O, Bigalke H, Dengler R, Wegner F, Degroot M, Wohlfarth K. Neutralizing antibodies and secondary therapy failure after treatment with botulinum toxin type A: Much ado about nothing? Clinical Neuropharmacology. 2009;**32**(4):213-218

[20] Bellows S, Jankovic J. Reply to comment on re-visiting immunogenicity associated with botulinum toxin treatment. Toxins. 2019;**11**:491 Toxins. 2020;12(2):72

[21] Hefter H, Hartmann C, Kahlen U, Moll M, Bigalke H. Prospective analysis of neutralising antibody titres in secondary non-responders under continuous treatment with a botulinumtoxin type A preparation free of complexing proteins—A single cohort 4-year follow-up study. BMJ Open. 2012;**2**(4):e000646

[22] Greene P, Fahn S, Diamond B. Development of resistance to botulinum toxin type A in patients with torticollis. Movement Disorders. 1994;**9**(2):213-217

[23] Jankovic J, Vuong KD, Ahsan J. Comparison of efficacy and immunogenicity of original versus current botulinum toxin in cervical dystonia. Neurology. 2003;**60**(7):1186-1188

[24] Brin MF, Comella CL, Jankovic J, Lai F, Naumann M. Long-term treatment with botulinum toxin type A in cervical dystonia has low immunogenicity by mouse protection assay. Movement Disorders. 2008;**23**(10):1353-1360

[25] Benecke R, Jost WH, Kanovsky P, Ruzicka E, Comes G, Grafe S. A new botulinum toxin type A free of complexing proteins for treatment of cervical dystonia. Neurology. 2005;**64**(11):1949-1951

[26] Frevert J. Pharmaceutical, biological, and clinical properties of botulinum neurotoxin type A products. Drugs R & D. 2015;**15**(1):1-9

[27] Moll M, Rosenthal D, Hefter H. Quality of life in long-term botulinum toxin treatment of cervical dystonia: Results of a cross sectional study. Parkinsonism & Related Disorders. 2018;**57**:63-67

[28] Hefter H, Rosenthal D, Bigalke H, Moll M. Clinical relevance of neutralizing antibodies in botulinum toxin long-term treated still-responding patients with cervical dystonia. Therapeutic Advances in Neurological Disorders. 2019;**12**:175628641 9892078